KNOWLEDGE COORDINATION PATTERNS

TWELVE HISTORICAL EXPERIMENTS IN INTERDISCIPLINARY COORDINATION

HAKIM IBN ADAM

Centre for Studies in Matter, Mind, and Meaning

threerosespublishing.com

ISBN: 978-1-9990656-7-6

AUTHOR'S NOTE

This book is a publication of the "Centre for Studies in Matter, Mind, and Meaning", a division of Three Roses Publishing. The Centre's mission is to produce scholarly works that integrate scientific inquiry, philosophical analysis, and spiritual reflection.

CONTENTS

Chapter 4: Medieval-Renaissance Coordination Experiments - Aquinas and Leonardo as Bounded Case Studies........................... 26

PREFACE

Contemporary knowledge production faces genuine coordination challenges. Specialized expertise has enabled remarkable achievements, yet this same specialization creates interface problems when complex issues require collaboration across domains. Translation costs increase as vocabularies diverge, institutional incentives often penalize boundary-crossing work, and format incompatibilities impede data sharing. These are not symptoms of intellectual crisis but predictable design constraints of successful knowledge systems.

The question is not whether specialization represents decline from some imagined golden age of unified wisdom, but where coordination failures genuinely impede progress on problems that matter. Climate change, pandemic response, and technological governance all require orchestrating insights across multiple domains without sacrificing the rigor that makes specialized knowledge valuable in the first place.

This book examines twelve historical figures who attempted various forms of knowledge synthesis across different cultural and institutional contexts. Rather than treating them as universal models or sources of timeless principles, the book approaches them as historically situated experiments in coordination. Each faced particular challenges within specific constraints—court patronage, religious orthodoxy, institutional limitations, or disciplinary boundaries of their era.

Our protagonists include Pythagoras and Plato from classical antiquity, Islamic scholars Avicenna and Al-Biruni, medieval synthesist Aquinas, and Renaissance polymath Leonardo, Enlightenment systematisers Leibniz and Goethe, and modern figures Teilhard, Iqbal, Jung, and Bohm. Their approaches ranged from mathematical mysticism through theological coordination to psychological and physical theories of wholeness. Some achieved durable coordination within their contexts; others produced ambitious failures that nonetheless illuminate persistent challenges.

The book analyzes these cases not to extract universal laws but to identify transferable practices, recurring obstacles, and boundary conditions for successful coordination. Four patterns emerge repeatedly across different contexts, which the book treats as working hypotheses rather than established principles: systematic coordination of different validation methods, hierarchical organization that preserves disciplinary integrity while enabling interface, process-oriented thinking that emphasizes relationships over static categories, and participatory approaches that acknowledge the researcher's situatedness.

These patterns deserve investigation precisely because they appear in such different contexts, but their apparent universality may reflect selection bias, cultural narrowness, or interpretive projection. The book therefore, establishes several methodological guardrails: distinguishing empirical claims from normative commitments, avoiding anachronistic attribution of contemporary concerns to historical figures, maintaining clear boundaries between different warrant domains, and explicitly examining who was included or excluded from these coordination projects.

Rather than seeking metaphysical unity or defending fragmentation, the book frames coordination as interface design—the creation of durable coordination mechanisms between heterogeneous knowledge communities. This includes developing shared boundary objects (datasets, standards, conceptual bridges), minimal ontologies that enable interoperation without forcing homogenization, incentive structures that reward integrative work, and governance frameworks that protect methodological diversity while enabling joint action.

This approach acknowledges that some domains may be irreducibly different, that forced synthesis often produces neither good science nor good philosophy, and that apparent unity in earlier historical periods frequently masked exclusions and power dynamics rather than genuine intellectual achievement. It also recognizes that effective coordination sometimes requires accepting productive tensions rather than resolving them prematurely.

This investigation has clear limitations. Twelve figures from primarily

Western and Islamic traditions cannot establish universal patterns. Many represent elite male perspectives from particular institutional contexts. Alternative knowledge traditions—Indigenous, African, East Asian, and others—receive insufficient attention. The selection of recognized "integrators" may create circular reasoning about coordination methods.

The book also brackets certain contemporary claims. While coordination problems are real, attributing widespread cultural anxieties primarily to academic specialization lacks robust evidence. Multiple social determinants—economic inequality, political instability, technological disruption—likely matter more than disciplinary boundaries for understanding contemporary psychological distress.

Success for this project is not rhetorical persuasion about the necessity of coordination, but clearer understanding of coordination challenges and a practical toolkit for addressing them. The book seeks design principles, interface exemplars, and boundary conditions that readers can adapt to their own integrative tasks in research, education, policy, and practice.

Where historical evidence supports claims about effective coordination methods, we document them carefully. Where it reveals the costs, failures, or exclusions required to maintain apparent unity, the book examines those honestly. Where different domains resist synthesis, acknowledging those limits is itself valuable knowledge.

The goal is neither nostalgic return to pre-modern unity nor uncritical celebration of fragmentation, but practical wisdom about when, how, and why to attempt coordination across knowledge domains—and when to respect productive differences instead.

CHAPTER 1: KNOWLEDGE COORDINATION IN AN ERA OF PRODUCTIVE SPECIALIZATION

ABSTRACT

This chapter examines contemporary knowledge coordination challenges while rejecting both crisis narratives and uncritical defenses of disciplinary isolation. The book defines fragmentation operationally as observable interface problems rather than as civilizational pathology, acknowledges specialization's remarkable achievements, and identifies specific coordination bottlenecks that impede societally important work. Rather than seeking metaphysical unity, pragmatic coordination focused on improving translation is approached, developing shared infrastructure, and designing incentives for productive boundary-crossing. Historical precedents inform this approach not as universal models but as contextually situated experiments offering transferable insights about coordination successes and failures.

1.1 REFRAMING THE COORDINATION CHALLENGE

Contemporary knowledge production exhibits both unprecedented depth within specialized domains and genuine challenges at disciplinary interfaces. Rather than characterizing this situation as a "crisis," we examine it as a predictable feature of mature knowledge systems that requires targeted coordination responses where interface problems impede important work.

Specialization has enabled remarkable achievements across virtually every domain of human inquiry. The development of mRNA vaccines drew on decades of focused research in molecular biology, immunology, and pharmaceutical chemistry (Karikó et al., 2005; Pardi, Hogan, Porter, &

Weissman, 2018). Climate science advances through coordination among atmospheric physicists, oceanographers, glaciologists, and ecological modelers—each contributing specialized expertise that no generalist could match. Cancer treatment improvements reflect molecular biology's detailed understanding of cellular mechanisms, not philosophical synthesis about the "unity of life" (Hanahan & Weinberg, 2011; Vogelstein et al., 2013).

These successes suggest that the question is not whether specialization is beneficial—the evidence strongly supports its value—but where coordination failures genuinely impede progress on problems that matter. Interface challenges become problematic when they prevent effective collaboration on multi-scale issues like pandemic response, climate adaptation, or technological governance that require orchestrating insights across multiple domains (Wuchty, Jones, & Uzzi, 2007; Eccleston-Turner & Upton, 2021).

This framework recognizes three levels of coordination challenges: technical interface problems (data standards, methodological differences), institutional-structural barriers (incentive misalignments, resource inequalities), and epistemic-political tensions (whose knowledge counts, which problems receive attention).

1.2 OPERATIONAL DEFINITION OF FRAGMENTATION

Rather than invoking abstract notions of knowledge "unity" or "crisis," we define fragmentation through specific, observable coordination problems:

Format and Ontological Incompatibilities: Different fields develop data standards, measurement protocols, and conceptual frameworks that resist coordination. Beyond technical mismatches, we must consider how different epistemic traditions conceptualize reality differently. Indigenous knowledge systems' holistic approaches to environmental understanding often cannot be "translated" into Western scientific categories without fundamental loss (Whyte, 2017).

Validation Norm Divergence: Different standards for evidence reflect not just methodological preferences but deeper questions about what constitutes legitimate knowledge. The exclusion of experiential, embodied, or narrative forms of knowing from many interdisciplinary collaborations represents an epistemic injustice that coordination efforts must address (Fricker, 2007; Wasserstein & Lazar, 2016; Geertz, 1973).

Incentive Misalignment: Academic reward structures frequently penalize boundary-crossing work (Leahey, Beckman, & Stanko, 2017; Bromham, Dinnage, & Hua, 2016). Interdisciplinary research faces challenges in peer review when reviewers lack expertise across relevant domains, and promotion committees may undervalue work that doesn't fit clearly within departmental boundaries.

Power Asymmetries: Coordination failures often reflect unequal power relations between fields, institutions, and knowledge communities. Economics' dominance in policy discussions exemplifies how disciplinary hierarchies impede genuine interdisciplinary dialogue (Fourcade, Ollion, & Algan, 2015; Nordhaus, 2017; Stern, 2007).

Translation Costs: Technical vocabularies and conceptual frameworks drift over time, increasing the effort required to communicate across fields (Bowker & Star, 1999; Star & Griesemer, 1989). What economists call "externalities," sociologists might term "unintended consequences," and systems theorists describe as "emergent properties"—conceptual differences that complicate collaboration even when substantive agreement exists (Anderson, 1972; S. D. Mitchell, 2009)

These operational definitions allow us to identify specific coordination bottlenecks without invoking grand narratives about the fragmentation of human understanding or the loss of civilizational wisdom.

1.3 HISTORICAL PERSPECTIVE: UNITY, CONSTRAINT, AND SYNTHESIS

Contemporary coordination challenges must be understood against appropriate historical baselines. Claims about a golden age of unified knowledge typically romanticize periods when apparent unity reflected institutional constraints rather than genuine coordination (Lindberg, 2007; Shapin, 1996).

Medieval "natural philosophy" exhibited apparent coherence partly because inquiry operated within narrow institutional boundaries, limited populations of scholars, and theological frameworks that predetermined acceptable conclusions. The transition to specialized sciences involved not just institutional reorganization but methodological liberation—the development of domain-specific investigative tools that enabled discoveries impossible under generalist natural philosophy.

Galileo's telescopic observations, Harvey's circulation experiments, and Newton's mathematical mechanics succeeded precisely by abandoning comprehensive philosophical systems in favor of focused empirical investigation. The subsequent proliferation of specialized methods reflects the discovery that different phenomena require different investigative approaches, not intellectual decline from some integrated ideal.

While medieval natural philosophy did operate under theological constraints, it also achieved genuine synthetic insights about the interconnection of natural phenomena that specialized sciences later struggled to recapture. Consider Ibn Rushd (Averroes) and Thomas Aquinas, whose coordination of Aristotelian philosophy with revealed theology represented sophisticated coordination between different knowledge traditions, not mere institutional constraint (McGrath, 2004). Similarly, Alexander von Humboldt's comprehensive natural philosophy, while impossible to replicate today, generated ecological insights about planetary interconnection that preceded and enabled later specialized discoveries (Wulf, 2015).

1.4 LEARNING FROM COORDINATION ATTEMPTS

Past efforts at large-scale knowledge coordination provide cautionary lessons about both possibilities and limitations of coordination projects.

Edward O. Wilson's Consilience (1998) proposed reducing all valid knowledge to fundamental physical laws through hierarchical explanatory chains. While appealing in its systematic ambition, this reductionist approach encountered fundamental obstacles. Emergent properties at higher organizational levels—from chemistry to biology to psychology—resist complete explanation through lower-level mechanisms. Water's properties cannot be predicted solely from hydrogen and oxygen; ecosystems exhibit dynamics invisible at the organism level; consciousness displays features absent from neural correlates (Levin, 1998; Chalmers, 1995).

While hierarchical reduction failed as a complete coordination strategy, Wilson's emphasis on gene–culture coevolution productively influenced fields from evolutionary psychology to cultural epidemiology (Henrich, 2016).

Systems theory and complexity science offered alternative coordination strategies by identifying patterns across domains—feedback loops, scaling relationships, network structures—without requiring reductive explanation (M. Mitchell, 2009; Page, 2011). The Santa Fe Institute's interdisciplinary approach generated valuable insights about emergence, adaptation, and collective behavior. However, these programs achieved limited institutional penetration, remaining marginal to mainstream disciplinary structures despite decades of development.

More recent "convergence science" initiatives show promise by focusing on specific problem domains rather than comprehensive theoretical unification. The National Science Foundation's Convergence Accelerator explicitly addresses coordination challenges through team-based approaches, shared infrastructure, and novel funding mechanisms (Sharp & Langer, 2011). Early assessments suggest greater success when coordination efforts target particular coordination bottlenecks rather than pursuing grand theoretical synthesis.

Non-Western coordination attempts offer underexplored models. Traditional Chinese medicine's systematic coordination of empirical observation with theoretical frameworks, while not meeting Western scientific standards, demonstrates alternative approaches to knowledge coordination that inform contemporary integrative medicine (Scheid, 2002).

1.5 CONTEMPORARY COORDINATION STRATEGIES

Effective coordination strategies emerge from successful inter-field collaborations rather than top-down theoretical frameworks. Several approaches show particular promise:

Boundary Objects and Shared Infrastructure: The most successful interdisciplinary collaborations develop artifacts that different fields can use without abandoning their methodological commitments (Star & Griesemer, 1989; Bojinski et al., 2014; Benson et al., 2018). Climate science's Global Climate Observing System provides standardized data that atmospheric physicists, ecologists, and policy analysts can all utilize while maintaining their distinct analytical approaches. Climate data standards, while enabling coordination, also embed particular assumptions about what matters to measure (Edwards, 2010). Similarly, genomic databases enable coordination between molecular biologists, evolutionary researchers, and medical practitioners without requiring methodological convergence.

Translational Layers: Effective coordination often requires explicit translation work—developing glossaries, concept maps, and interface protocols that enable communication across different technical vocabularies (Beck & Mahony, 2017; Mach, Mastrandrea, Freeman, & Field, 2017). The

Intergovernmental Panel on Climate Change (IPCC) succeeds partly through systematic translation between scientific technical language and policy-relevant communication, maintaining scientific rigor while enabling practical application. The IPCC's translation between scientific and policy languages shapes both domains, not just communicating neutral facts (Hulme, 2009; Chakrabarty, 2009).

Incentive Design with Equity Focus: Institutional innovations can address coordination problems by recognizing boundary-spanning contributions. Some universities now evaluate interdisciplinary work through mixed-field review panels, while funding agencies increasingly value broader impacts alongside disciplinary contributions. These developments suggest that coordination problems often reflect institutional design choices rather than fundamental incompatibilities between domains. New incentive structures must address not just interdisciplinarity but also epistemic diversity and inclusion of marginalized knowledge communities. Decolonizing methodologies offer frameworks for more equitable coordination (Smith, 2012; Mirowski, 2011).

Methodological Pluralism with Explicit Scope Conditions: Rather than seeking universal methods, successful coordination projects often explicitly map different approaches' scope conditions—where each method works well, where it encounters limits, and how different methods can complement rather than compete (S. D. Mitchell, 2009; Kellert, Longino, & Waters, 2006). This approach preserves the integrity of specialized expertise while enabling productive collaboration.

Epistemic Humility and Power Sharing: Effective coordination requires dominant fields to cede epistemic authority and share resources with marginalized knowledge traditions (Harding, 2015).

1.6 COORDINATION WITHOUT UNIFICATION

The distinction between coordination and unification proves crucial for realistic coordination projects. Coordination aims to improve interfaces between heterogeneous approaches without eliminating their differences. Unification seeks a deeper theoretical synthesis that reduces multiple approaches to common foundations.

While unification remains a legitimate long-term research program, coordination offers more immediate practical value. Climate policy requires coordinating insights from atmospheric physics, economics, political science, and engineering without first achieving theoretical unification across these domains (Kitcher, 1989). Similarly, pandemic response benefits from coordination among virology, immunology, epidemiology, and public health policy, regardless of whether these fields share common foundational assumptions. Civilizational challenges like climate change and Anthropocene-scale transformation may require selective theoretical synthesis in specific domains while maintaining methodological pluralism elsewhere (Chakrabarty, 2009).

This pragmatic approach acknowledges that some domains may be irreducibly different—requiring distinct methods, validation criteria, and conceptual frameworks that resist synthesis. Rather than treating such differences as problems to be solved, coordination approaches them as design constraints to be managed through improved interface mechanisms. Coordination suffices for many practical problems, but civilizational challenges like climate change, artificial intelligence governance, and biotechnology ethics may require selective theoretical synthesis in specific domains while maintaining methodological pluralism elsewhere.

1.7 SCOPE CONDITIONS AND LIMITATIONS

Any realistic coordination agenda must acknowledge clear scope conditions and limitations. Not all coordination problems are worth solving, and some may be unsolvable given current institutional and cognitive constraints.

Cost-Benefit Analysis: Coordination efforts require significant time and resource investments. These costs are justified when coordination failures genuinely impede important work, but not for purely academic coordination exercises. The burden of proof lies on coordination advocates to demonstrate specific benefits that justify coordination costs.

Productive Tensions: Some disciplinary differences generate creative friction that drives research progress. Competing methodological approaches, theoretical frameworks, and interpretive traditions may produce better knowledge through productive disagreement than through premature synthesis. Forced coordination risks eliminating tensions that stimulate discovery.

Institutional Realities: Coordination improvements must work within existing institutional structures—universities, funding agencies, journals, professional organizations—that have evolved to support specialized research. Proposals that require wholesale institutional reorganization face practical obstacles that limit their viability regardless of their theoretical merits. However, crisis moments (like pandemics) can create windows for institutional innovation. The rapid establishment of COVID research collaborations demonstrates latent institutional flexibility (Giles, 2020; Eccleston-Turner & Upton, 2021; Mirowski, 2011).

Cognitive Limitations: Individual scholars have finite capacity for mastering multiple domains deeply. Effective coordination often requires teams with complementary expertise rather than polymaths who master everything personally (Wuchty, Jones, & Uzzi, 2007). This suggests that coordination solutions should emphasize collaborative mechanisms rather than individual synthetic capacity. However, collective intelligence approaches (including AI-assisted synthesis) may expand coordination possibilities beyond current assumptions (Malone, 2018).

Justice and Inclusion Constraints: Coordination efforts that exclude marginalized communities or reinforce existing inequalities are ethically problematic regardless of technical efficiency. Environmental justice movements demonstrate how community knowledge must inform scientific coordination (Bullard & Wright, 2009).

1.8 IMPLICATIONS FOR HISTORICAL ANALYSIS

These considerations establish methodological principles for analyzing historical coordination attempts in subsequent chapters. Rather than seeking universal principles or timeless models, we examine each case as a situated experiment in coordination under particular constraints.

For each historical figure, we ask: What specific coordination challenges did they face within their institutional and cultural context? Which coordination strategies proved effective for their particular problems? What resources—institutional, conceptual, social—enabled their integrative work? What exclusions or limitations shaped their apparent successes? Which aspects of their approach might transfer to contemporary coordination challenges, and which remain historically specific?

This approach treats historical integrators neither as heroes to emulate nor as cautionary tales to avoid, but as natural experiments in coordination that can inform contemporary efforts while respecting the particularity of their contexts and constraints.

1.9 ROADMAP FOR THE INVESTIGATION

The chapters that follow examine twelve historical figures who attempted various forms of knowledge coordination across different periods and contexts. Rather than demonstrating universal coordination principles, these cases illustrate the diversity of coordination approaches, the contextual factors that enable or constrain integrative work, and the transferable insights that survive translation across historical contexts.

Each case study examines both successes and limitations, attending particularly to what coordination problems each figure addressed, what resources enabled their approach, what constraints shaped their work, and what aspects of their methods might inform contemporary coordination efforts. The goal is not to celebrate coordination as inherently valuable, but to understand when, how, and why coordination efforts succeed or fail.

The investigation concludes not with universal prescriptions for knowledge coordination, but with a toolkit of coordination strategies, scope conditions for their application, and realistic assessments of what coordination can and cannot accomplish in contemporary knowledge production.

CHAPTER 2: PYTHAGORAS AND PLATO — MATHEMATICAL STRUCTURE AS COORDINATION METHOD

ABSTRACT

This chapter examines early Greek attempts to coordinate knowledge through mathematical structure, focusing on Pythagorean traditions and Platonic dialectical methodology. Rather than treating these figures as universal coordination models, we analyze them as historically situated experiments in using formal patterns to organize inquiry across domains. While acknowledging serious limitations—legendary source material for Pythagoras, authoritarian implications in Plato's political philosophy, and systematic gender exclusions—we extract methodologically useful insights about structural thinking, hierarchical organization, and dialectical method. The analysis maintains clear boundaries between ancient achievements and contemporary applications, avoiding anachronistic projections while identifying transferable coordination principles.

2.1 HISTORICAL CONTEXT: FRAGMENTATION AS INTELLECTUAL CHALLENGE

Sixth-century BCE Greece confronted genuine coordination challenges as competing explanatory systems proliferated without clear adjudication mechanisms (Lloyd, 2021; McKirahan, 2020). Milesian naturalists, Heraclitean flux theorists, and Eleatic monists proposed incompatible foundational principles—water, fire, and unchanging being respectively—creating what G. E. R. Lloyd terms an "epistemological crisis," where unified understanding appeared threatened by irreducible theoretical pluralism (Lloyd, 2021).

This context shaped distinctive Greek responses to coordination challenges. Rather than accepting theoretical fragmentation or imposing institutional orthodoxy, certain traditions developed methodological approaches for identifying formal patterns that could organize inquiry across different domains. These approaches merit examination not as successful solutions to contemporary coordination problems, but as early experiments in using mathematical structure and dialectical method as coordination mechanisms.

2.2 PYTHAGOREAN TRADITIONS: PATTERN RECOGNITION AND ITS LIMITS

The historical Pythagoras remains largely inaccessible through reliable sources (Huffman, 2019; Zhmud, 2022). No writings survive, and our primary evidence comes from Philolaus (late 5th century), Aristotle's references to "so-called Pythagoreans," and much later Neopythagorean reconstructions (Huffman, 2019; Zhmud, 2022). This source problem requires attributing ideas cautiously to "Pythagorean traditions" rather than to Pythagoras himself, and distinguishing between early mathematical insights and later mystical elaborations.

2.2.1 Mathematical Discoveries and Methodological Significance

A genuine achievement of early Pythagorean traditions appears to be recognizing that quantitative relationships could illuminate qualitative phenomena across different domains. The discovery that musical harmony corresponds to simple whole-number ratios—octaves (2:1), fifths (3:2), fourths (4:3)—represents what Walter Burkert identifies as "the first time in history that a mathematical formula was found to describe a physical phenomenon accurately" (Burkert, 2020; McKirahan, 2020).

This breakthrough established a methodological precedent: formal relationships discovered in one domain might illuminate patterns in others. Applied systematically, this approach suggested that mathematical structure rather than material substance might constitute reality's fundamental level of organization. As Aristotle reports, Pythagoreans "saw many resemblances in numbers to the things that exist and come into being—more than in fire and earth and water" (Metaphysics 985b23-26) (Aristotle, 1984).

2.2.2 Coordination Through Proportion

The tetractys—triangular arrangement of ten points in four rows (1+2+3+4=10)—functioned as a coordination device encoding musical intervals, geometric relationships, and cosmological principles within a unified structure (Burkert, 2020). Rather than accepting this as sacred revelation, we can understand it as an early attempt at cross-domain pattern recognition: identifying formal structures that appear in multiple contexts and using them to coordinate understanding across those contexts.

This approach enabled certain coordination successes within its historical context. Musical theory, mathematical geometry, and ethical reflection could be organized through proportional relationships, creating shared vocabulary and conceptual bridges between domains that might otherwise remain isolated.

2.2.3 Structural Limitations and Exclusions

However, this coordination came at significant costs that must be acknowledged. The Table of Opposites, preserved in Aristotle (Metaphysics 986a22-26), encoded hierarchical valuations—male/female, good/evil, light/darkness—that systematically excluded women and associated femininity with inferiority and irrationality (Aristotle, 1984; Zhmud, 2022). This exclusion was not incidental but constitutive of the coordination system, which achieved apparent unity partly by marginalizing perspectives that might challenge its harmonic vision.

The discovery of incommensurable quantities ($\sqrt{2}$) created a crisis for ratio-based cosmology, revealing that mathematical reality exceeded simple whole-number relationships (McKirahan, 2020). This forced recognition that the coordination method had scope limitations, though it also motivated mathematical generalization beyond elementary proportional thinking.

2.3 PLATONIC DIALECTICAL ARCHITECTURE

Plato transformed Pythagorean insights about mathematical structure into systematic philosophical methodology through dialectical investigation and the theory of Forms (Plato, 1997c). Rather than treating Plato's philosophy as a universal coordination model, we examine specific methodological contributions that remain relevant for contemporary coordination challenges.

2.3.1 The Divided Line as Organizational Framework

The divided-line passage (Republic 509d-511e) provides not metaphysical doctrine but a methodological scaffold for organizing different types of cognitive engagement with reality (Plato, 1997c). The four levels— imagination (eikasia), belief (pistis), understanding (dianoia), and knowledge (noēsis)—establish relationships between different warrant domains without eliminating their distinctiveness (Plato, 1997c).

This framework enables coordination by clarifying how different approaches to inquiry relate hierarchically while maintaining their autonomous validity. Empirical observation, mathematical reasoning, and philosophical dialectic each contribute legitimately to understanding without requiring reduction to a single method. The coordination is achieved through explicit specification of relationships rather than through theoretical unification.

2.3.2 Dialectical Method as Coordination Tool

The method of division and collection (diairesis kai synagōgē) developed in dialogues like the Sophist and the Philebus, and the programmatic praise of such a method in the Phaedrus, provides systematic procedures for identifying both differences and commonalities across domains (Plato, 1997d; 1997a; 1997b). As Plato has Socrates say: "I am myself a lover of these divisions and collections, that I may be able to speak and think" (Phaedrus 266b3-4) (Plato, 1997b).

This methodology coordinates by making explicit the conceptual work required to move between different levels of generality and specificity. Rather than assuming unity, the dialectical method investigates where unity exists, where differences persist, and how both can be systematically mapped. This approach provides transferable coordination principles: explicit articulation of relationships, systematic attention to scope conditions, and preservation of relevant differences alongside identification of genuine commonalities.

2.3.3 Forms Theory: Strengths and Limitations

The theory of Forms attempts to explain why mathematical and conceptual coordination works by positing transcendent structures that particular phenomena instantiate (Plato, 1997c). While this metaphysical framework faces well-known objections—the "third man" regress, participation puzzles, and lack of independent empirical support—it contains methodologically valuable insights about structural thinking (Fine, 2019).

The emphasis on formal organization rather than material substrate anticipates contemporary recognition that information and pattern, rather than substance, may constitute fundamental explanatory categories (French, 2023; Mitchell, 2009). However, treating this as an empirical vindication of Platonic metaphysics would be anachronistic overreach. The value lies in the methodological focus on structure and relationship rather than in specific metaphysical commitments.

2.4 CROSS-DOMAIN APPLICATIONS: ACHIEVEMENTS AND CONSTRAINTS

Both Pythagorean and Platonic approaches achieved coordination successes within their historical contexts by applying formal principles across cosmology, psychology, ethics, and politics. The Timaeus presents a cosmos structured by mathematical principles, while the Republic applies tripartite soul analysis to political organization (Plato, 1997e; 1997c). These demonstrate systematic application of coordination principles beyond their original domains.

However, these applications operated within significant constraints that limit their contemporary relevance. Plato's political philosophy in the Republic includes authoritarian elements—philosopher-kings with absolute authority, noble lies, censorship—that Karl Popper influentially criticized as proto-totalitarian (Popper, 2011). The apparent coordination was achieved partly through suppressing dissent and diversity rather than through genuine coordination of different perspectives.

Gender exclusions remained systematic throughout these traditions despite occasional gestures toward inclusion. While Republic Book V suggests women might serve as guardians, other dialogues (e.g., Timaeus) maintain traditional restrictions, and the historical Academy apparently included no women members (Plato, 1997e; Lloyd, 2021). This exclusion shaped the coordination systems fundamentally, not peripherally.

2.5 CONTEMPORARY PARALLELS: BOUNDED ANALOGIES

Certain structural features of ancient Greek mathematical approaches exhibit limited parallels to contemporary coordination methods in network theory, complexity science, and information theory (Barabási, 2016;

Mitchell, 2009; Thurner, Hanel, & Klimek, 2018). However, these analogies must be stated with explicit scope conditions to avoid anachronistic projection.

2.5.1 Structure and Relationship

The Pythagorean recognition that relationships rather than substances might constitute fundamental reality anticipates contemporary structural approaches in physics and mathematics. Network theory similarly focuses on patterns of connection rather than node properties, while information theory treats structure and pattern as fundamental explanatory categories.

However, ancient Greeks lacked the mathematical tools—set theory, topology, computational methods—that enable contemporary network analysis. The parallel lies in methodological orientation toward structure, not in specific theoretical content.

2.5.2 Hierarchical Organization

Plato's divided line provides early systematic thinking about how different organizational levels relate without reduction. This anticipates contemporary multi-level approaches in complexity science that recognize emergent properties while maintaining connections across scales.

But ancient hierarchical thinking often encoded value judgments (higher = better) that contemporary complexity science explicitly avoids. The methodological insight about multi-level organization can be preserved while rejecting evaluative hierarchies that marginalize certain approaches or perspectives.

2.5.3 Pattern Recognition Across Domains

Both traditions demonstrated systematic attention to formal patterns that appear in multiple contexts—proportional relationships, geometric structures, organizational principles. This anticipates contemporary recognition of scaling laws, network patterns, and information-theoretic principles that appear across different systems (West, 2017).

However, contemporary pattern recognition benefits from statistical methods, computational analysis, and large datasets unavailable to ancient thinkers. The parallel lies in the methodological commitment to systematic pattern identification, not in the specific patterns identified.

2.6 METHODOLOGICAL LESSONS FOR CONTEMPORARY COORDINATION

Extracting from these historical cases while avoiding their limitations suggests several methodological principles for contemporary coordination efforts:

Structural Thinking: Focus on relationships, patterns, and organizational principles rather than seeking substantive theoretical unity. This enables coordination across domains with different methodological commitments.

Explicit Hierarchy: When hierarchical organization is useful, make the organizational principles explicit and separate them from value judgments. Specify clearly what "higher" and "lower" mean operationally.

Dialectical Investigation: Before assuming unity or difference, investigate systematically where each exists and under what conditions. Use explicit procedures for mapping relationships rather than asserting them.

Inclusion Audit: Examine systematically who is authorized to contribute to coordination efforts and how exclusions shape apparent unity. Design coordination mechanisms to include rather than marginalize different perspectives.

Scope Limitation: Acknowledge explicitly where coordination methods work well and where they encounter limitations. Avoid extending successful coordination beyond appropriate scope conditions.

2.7 ISLAMIC TRANSMISSION AND SELECTIVE APPROPRIATION

The transmission of Pythagorean-Platonic coordination methods through Islamic philosophy demonstrates how coordination approaches can be selectively appropriated and adapted across cultural contexts. Figures like al-Kindī and the Brethren of Purity preserved mathematical approaches to coordination while modifying metaphysical commitments to fit Islamic theological contexts (Adamson, 2016).

This selective appropriation suggests that coordination methods may transfer more successfully than the comprehensive theoretical frameworks within which they originally developed. Islamic philosophers extracted useful methodological tools—numerical pattern recognition, hierarchical organization, dialectical investigation—while rejecting or modifying metaphysical commitments that conflicted with Islamic theology.

This provides a model for contemporary appropriation: identify transferable coordination methods while explicitly bracketing metaphysical frameworks that may not transfer across different contexts and commitments.

2.8 CRITICAL ASSESSMENT AND HISTORICAL LIMITATIONS

Several important limitations constrain the contemporary relevance of these ancient coordination approaches:

Source Problems: For Pythagorean traditions, legendary attributions and late sources make it difficult to distinguish genuine early insights from later reconstructions. This requires careful attention to what can be reliably attributed to historical figures versus traditional ascriptions.

Institutional Constraints: Both traditions operated within small, elite communities with shared cultural assumptions. Their coordination methods may not scale to contemporary institutional contexts with greater diversity and democratic accountability requirements.

Exclusionary Practices: The systematic exclusion of women, slaves, and non-Greeks shaped these coordination systems fundamentally. Contemporary applications must address rather than reproduce these exclusions.

Methodological Limitations: Ancient thinkers lacked empirical methods, statistical analysis, and computational tools that enable contemporary coordination approaches. Their insights about structure and pattern must be supplemented with more rigorous investigative methods.

Authoritarian Implications: Platonic political philosophy contains elements that conflict with democratic values and pluralistic coordination. The philosophical-king model cannot be transferred uncritically to contemporary coordination challenges.

2.9 DESIGN PRINCIPLES FOR MATHEMATICAL COORDINATION

Despite these limitations, several design principles emerge from successful aspects of ancient Greek mathematical coordination:

Pattern Before Unity: Identify formal patterns systematically before asserting deeper theoretical unity. This enables coordination without premature theoretical commitment.

Method Before Metaphysics: Develop coordination methods that can work across different metaphysical frameworks rather than requiring agreement on fundamental theoretical assumptions.

Structure Before Substance: Focus on organizational relationships and formal properties rather than seeking agreement on underlying substantive commitments.

Hierarchy with Inclusion: When hierarchical organization proves useful, design it to include rather than exclude different approaches and perspectives.

Explicit Scope Conditions: Specify clearly where coordination methods work well and where they encounter limitations. Avoid extending successful coordination beyond appropriate domains.

2.10 CONCLUSION: METHODOLOGICAL HERITAGE AND CONTEMPORARY APPLICATION

Ancient Greek mathematical approaches to coordination provide methodologically useful insights while requiring careful critical assessment. The focus on structural relationships, hierarchical organization, and systematic dialectical investigation offers transferable coordination principles that remain relevant for contemporary challenges.

However, these insights must be extracted from their original contexts while avoiding their exclusionary practices, authoritarian implications, and metaphysical overreach. The goal is not to emulate ancient Greek coordination systems wholesale, but to identify methodological principles that can inform contemporary coordination efforts within very different institutional and cultural contexts.

The historical analysis demonstrates both the possibilities and limitations of mathematical coordination approaches. They can enable productive collaboration across different domains and perspectives when properly designed, but they can also mask power relations, exclude alternative viewpoints, and impose premature theoretical unity. Contemporary applications must learn from both the successes and failures of these early coordination experiments.

CHAPTER 3: TRANSLATION AS COORDINATION — AVICENNA AND AL-BIRUNI IN HISTORICAL CONTEXT

ABSTRACT

This chapter examines two medieval Islamic scholars as case studies in coordination across cultural and disciplinary boundaries: Avicenna's systematic philosophical architecture and al-Biruni's empirical-comparative methodology. Rather than treating the medieval Islamic world as a "Golden Age" of seamless coordination, we analyze concrete coordination innovations within their institutional and political contexts. We extract transferable principles—creative terminology, hierarchical organization with explicit bridges, empirical cross-cultural comparison—while acknowledging limitations including patronage constraints, exclusions, and incomplete syntheses.

3. 1 HISTORICAL CONTEXT: PATRONAGE NETWORKS AND KNOWLEDGE PRODUCTION

The Abbasid translation movement (8th–11th centuries) created coordination opportunities through political expansion, economic resources, and cultural encounters, but operated within constraints that shaped intellectual production (Gutas, 2022; Saliba, 2007). The establishment of Baghdad in 762 CE positioned the Abbasid caliphate at the intersection of Persian administrative traditions, Byzantine Greek learning, and trade networks reaching India and Central Asia.

The Bayt al-Ḥikma (House of Wisdom) appears to have functioned chiefly as a court library and translation bureau rather than an independent research institution (Gutas, 2022). Archival analysis indicates that most scholarly work proceeded through dispersed patronage networks—private libraries, hospital-medical schools, observatories, and scholarly circles—rather than a centralized academy (Gutas, 2022).

3. 1. 1 The Translation Economy

Translation work operated within a competitive court economy where political legitimacy, administrative efficiency, and intellectual prestige intersected (Gutas, 2022). Abbasid sponsorship served to establish cultural authority distinct from Umayyad precedents and to compete with Byzantine claims to Greek heritage. This created opportunities but also constraints on which texts were translated and how they were interpreted.

Hunayn ibn Isḥāq (809–873) articulated principles distinguishing literal rendering from meaning-based translation, enabling conceptual innovation through creative terminology (Lamoreaux, 2016; Gutas, 2022). Teams of translators, revisers, and commentators collaborated to build an Arabic philosophical lexicon that enabled novel theoretical developments.

Selectivity marked every stage of the process. Texts were chosen for perceived utility to theological, legal, and administrative purposes. Greek skepticism, some materialist currents, and political theory incompatible with monarchical rule were largely excluded. Selective appropriation created possibilities for synthesis while foreclosing others.

3. 1. 2 Social Constraints and Exclusions

Knowledge production operated within hierarchical social structures that systematically excluded certain groups while including others. While individual women like Fāṭima al-Fihri (founder associated with al-Qarawiyyīn) or Sutayta al-Mahāmilī (mathematician) achieved recognition, these were exceptions often requiring unusual family connections or court patronage (Makdisi, 1981; Pormann & Savage-Smith, 2021).

Much translation and technical work was performed by religious minorities—Christians, Jews, and Zoroastrians—often of servile or recently manumitted status (Gutas, 2022). These contributions enabled coordination achievements we associate with Islamic scholarship, but social conditions constrained full participation in philosophical synthesis versus technical labor. Apparent unity often reflected limited participation rather than genuine coordination of diverse perspectives.

3. 2 AVICENNA (IBN SĪNĀ): SYSTEMATIC COORDINATION THROUGH PHILOSOPHICAL ARCHITECTURE

Abū ʿAlī al-Ḥusayn ibn ʿAbd Allāh ibn Sīnā (980–1037) developed methods for coordinating Aristotelian logic, Neoplatonic metaphysics, and Islamic theology that created new philosophical possibilities while operating within institutional constraints.

3. 2. 1 The Essence–Existence Distinction as a Coordination Tool

Avicenna's influential distinction between essence (māhiyya) and existence (wujūd) enabled coordination between logical analysis and metaphysical commitment (Lizzini, 2021; Janos, 2020). In the Metaphysics of the *Healing*, he explains, in effect, that what a thing is (its quiddity) can be analyzed independently of whether it exists; existence is received and thus distinct in account (McGinnis, 2020; Lizzini, 2021). This permitted systematic treatment of how particulars relate to general categories without reduction to either Platonic Forms or Aristotelian substantial forms.

Limitations followed. The distinction raised questions about whether existence is a real property and how essences relate to existents. Later thinkers like Averroes criticized the approach as generating pseudo-problems (Averroes, 1954). Scholastic debates in Latin Europe extended these issues for centuries (Van den Bergh, 1954; Davidson, 1992).

3. 2. 2 Hierarchical Organization and an Emanationist Framework

Avicenna's emanationist cosmology coordinated divine simplicity with cosmic complexity through hierarchical principles preserving both transcendence and systematic connection. The Necessary Existent emanates existence through successive Intelligences, structuring relationships across metaphysical, cosmological, and psychological domains (Davidson, 1992; McGinnis, 2020).

The same architectural sensibility appears in the *Canon*: organization from general principles through simples to complex diseases and compound treatments coordinated theoretical foundations with clinical practice, enabling durable transfer across cultures (Pormann & Savage-Smith, 2021). Theological controversy followed: critics like al-Ghazālī argued that emanationist necessity conflicted with divine freedom and providence

(Griffel, 2009).

3. 2. 3 Methodological Achievements and Limits

Avicenna's achievement lay in procedures for moving between abstraction and application—logical analysis, metaphysical commitment, and clinical practice—through explicit bridges rather than reductive unification. The approach worked best within domains sharing monotheist assumptions; it was less successful across genuinely incommensurable frameworks.

3. 3 AL-BĪRŪNĪ (ABŪ RAYḤĀN AL-BĪRŪNĪ): EMPIRICAL–COMPARATIVE COORDINATION

Abū Rayḥān Muḥammad ibn Aḥmad al-Bīrūnī (973–1048) emphasized linguistic immersion, mathematical precision, and systematic comparison across cultural boundaries. His method anticipated modern comparative work while operating within medieval constraints.

3. 3. 1 Linguistic Immersion and Cultural Translation

Al-Bīrūnī learned Sanskrit, studied with Brahmin scholars, and translated texts to develop "understanding from within" before systematic comparison. In *India*, he notes his efforts to master language and sources to show that Indian scholars "possess a rational method, though it differs from ours" (Sachau, 1910).

His access was limited: focus on Brahmanical texts, little engagement with Buddhist/Jain sources, and constraints shaped by court networks and available interlocutors. Comparative analysis ultimately proceeded through Islamic categories, setting scope limits on cross-cultural understanding.

3. 3. 2 Mathematical Coordination Across Traditions

Al-Bīrūnī used mathematics as a coordination language bridging cultural differences. The *Canon Masudicus* synchronized calendrical systems— Greek, Persian, Indian, Hebrew, Christian, Islamic—via astronomical calculation (Saliba, 2007). Empirical measurements (e. g. , Earth's radius by horizon observation) provided validation methods that operated across traditions, demonstrating how systematic observation could coordinate diverse practices. He recognized that different parameters often reflect alternative observational lineages rather than error (Sachau, 1910).

3. 3. 3 Comparative Matrices and Incommensurability

Rather than forcing one-to-one translations, al-Bīrūnī developed comparative matrices that preserved differences while enabling productive comparison. He acknowledged deep divergences—"Hindus differ from us in everything…"—and devised procedures to compare without collapsing distinct frameworks. These methods worked best for mathematics and observation; religious/metaphysical frameworks remained more resistant to coordination.

3. 4 COORDINATION ACHIEVEMENTS AND INSTITUTIONAL CONSTRAINTS

3. 4. 1 Terminological Innovation

Arabic philosophical vocabulary—*wujūd, māhiyya, mumkin al-wujūd, wājib al-wujūd*—enabled distinctions that influenced both Islamic and European thought. Terminological coordination created shared resources operating across frameworks. Yet translation entailed conceptual transformation: *falsafa* differs from Greek *philosophia*; *wujūd* spans semantic ranges neither Greek *ousia* nor Latin *esse* capture (Gutas, 2022).

3. 4. 2 Hierarchical Organization Methods

Both scholars developed hierarchical organization coordinating domains while preserving relative autonomy: Avicenna's metaphysical levels; al-Bīrūnī's comparative classifications. These frameworks clarified relations between theory and practice, abstract and concrete, general and particular. But hierarchies often encoded value judgments privileging "higher" forms— abstract over practical, Islamic over non-Islamic—which constrained genuine coordination.

3. 5 CONTEMPORARY RELEVANCE: BOUNDED LESSONS FOR COORDINATION

3. 5. 1 Translation as Creative Coordination

Productive coordination across boundaries requires creative translation that transforms both source and target frameworks. Terminology-building, conceptual adaptation, and explicit bridge principles outperform literalism or enforced synthesis. Contemporary international collaborations face similar challenges. Medieval patronage/theology differed markedly from modern democratic/secular institutions (Makdisi, 1981); methods need adaptation, not wholesale transfer.

3. 5. 2 Empirical Methods for Cross-Cultural Understanding

Al-Bīrūnī's combination—immersion, measurement, comparison—offers tools for today's global collaborations (e. g. , coordinating Indigenous knowledge with atmospheric modeling). Mathematical mediation plus cultural immersion and systematic comparison enable collaboration without forced unification.

3. 5. 3 Systematic Organization Without Rigid Hierarchy

Both scholars modeled movement between theory/practice, abstract/concrete, general/particular. Their organizational methods can aid interdisciplinary work today—if stripped of evaluative hierarchies that subordinate certain approaches.

3. 6 CRITICAL ASSESSMENT: ACHIEVEMENTS WITHIN CONSTRAINTS

3. 6. 1 Selective Appropriation Rather Than Comprehensive Synthesis

Both engaged in selective appropriation guided by theological commitments, patronage requirements, and institutional resources. Apparent "synthesis" often involved creative reinterpretation. The misattributed *Theology of Aristotle* (Plotinian) facilitated innovation while embedding misunderstandings (Adamson, 2016; Gutas, 2022). Contemporary coordination should similarly expect selectivity and transformation rather than universal unification.

3. 6. 2 Institutional Constraints and Political Contexts

Court patronage provided resources while constraining research directions (Gutas, 2022; Saliba, 2007). Legitimacy requirements and orthodoxy pressures shaped viable projects. Analogously, modern funding and disciplinary incentives channel coordination today.

3. 6. 3 Exclusions as Constitutive Rather Than Incidental

Systematic exclusions (gender, status, religion) shaped coordination forms. Addressing such patterns requires institutional change as well as method. Contemporary applications must adapt historical methods to inclusive contexts.

3. 7 METHODOLOGICAL PRINCIPLES FOR CONTEMPORARY APPLICATION

• Creative translation: collaboratively invent shared terminology; aim for functional coordination, not theoretical unity.

• Hierarchical organization with explicit bridges: relate levels of analysis while avoiding implicit value hierarchies.

• Mathematical mediation where appropriate, with clear scope limits.

• Systematic comparison: preserve differences while enabling dialogue.

• Institutional pluralism: support multiple coordination venues and approaches.

• Inclusion audit: examine participation and how exclusions shape apparent unity.

• Scope limitation: specify domains where methods work and where they fail.

3. 8 CONCLUSION: TRANSLATION AS TRANSFORMATION

Avicenna and al-Bīrūnī represent sophisticated experiments in coordination across cultural and disciplinary boundaries within particular historical constraints. Their achievements—terminological innovation, hierarchical organization, empirical comparison—offer methodological resources for contemporary coordination, while their limitations illuminate persistent difficulties in genuine cross-cultural and interdisciplinary coordination. Applications should extract methodological insights while addressing the exclusionary and hierarchical aspects of medieval approaches, aiming at coordination that serves democratic and inclusive values.

CHAPTER 4: MEDIEVAL-RENAISSANCE COORDINATION EXPERIMENTS - AQUINAS AND LEONARDO AS BOUNDED CASE STUDIES

ABSTRACT

This chapter examines Thomas Aquinas (1225-1274) and Leonardo da Vinci (1452-1519) as experiments in coordinating different domains of inquiry within particular institutional and cultural constraints. Rather than treating them as successful universal coordination models, we analyze specific coordination methods they developed, the institutional contexts that enabled and constrained their work, and the scope limitations of their approaches. Aquinas developed systematic procedures for coordinating philosophical reasoning with theological commitment through analogical thinking and hierarchical organization. Leonardo pioneered empirical-aesthetic coordination through visual epistemology and biomimetic investigation (Kemp, 2019). Both achievements occurred within significant constraints—ecclesial orthodoxy, patronage requirements, systematic exclusions—that shaped their coordination possibilities. We extract methodological insights while acknowledging their historical specificity and avoiding anachronistic appropriation.

4.1 HISTORICAL AND INSTITUTIONAL CONTEXT

The coordination challenges Aquinas and Leonardo faced emerged from specific institutional and cultural contexts that both enabled and constrained their integrative work. Understanding these contexts proves crucial for assessing what they achieved and what contemporary applications might be appropriate.

4.1.1 Medieval University and Ecclesial Constraints

Aquinas worked within the emerging university system, particularly at the University of Paris, which represented new institutional arrangements for advanced learning (Grant, 1996). However, these universities operated under ecclesial authority with clear doctrinal boundaries (Grant, 1996). The Faculty of Theology's supremacy over other faculties meant that philosophical investigation, while encouraged, could not challenge fundamental Christian teachings.

The Dominican Order's educational approach, exemplified by Albert the Great's coordination of Aristotelian philosophy with empirical observation, provided methodological precedents for Aquinas's coordination work (Mulchahey, 2019). However, this coordination occurred within predetermined theological frameworks rather than through open-ended inquiry.

The rapid reception and subsequent controversy surrounding Aquinas's work illustrates these institutional constraints. The 1277 Condemnations by Bishop Tempier of Paris, targeting 219 propositions including several associated with Thomistic positions, demonstrate that his coordination approach was seen as potentially dangerous by ecclesiastical authorities (Grant, 1996). This suggests that his "synthesis" operated at the edge of institutional tolerance rather than representing settled consensus.

4.1.2 Renaissance Workshop Culture and Patronage

Leonardo operated within very different institutional contexts—Renaissance workshops and court patronage systems. The Florentine workshop system, exemplified by Verrocchio's bottega where Leonardo trained, integrated artistic, technical, and theoretical learning through collaborative practice rather than systematic disputation (Kemp, 2019; Baxandall, 1972).

Court patronage, particularly at the Sforza court in Milan, created opportunities for wide-ranging investigation while imposing practical constraints (Kemp, 2019). Patrons required military engineering, architectural design, and spectacular entertainment rather than systematic theoretical synthesis (Kemp, 2019). This context enabled Leonardo's empirical investigations while constraining their systematic development and dissemination.

The temporal gap between Aquinas and Leonardo (nearly 200 years) and their different institutional contexts complicate claims about methodological continuity. Their coordination approaches responded to different problems within different constraints rather than representing a unified tradition.

4.1.3 Gender and Social Exclusions as Structural Features

Both institutional contexts systematically excluded women and other groups from full participation in coordinated inquiry. Medieval universities prohibited women from formal study, while Renaissance workshops restricted their access to anatomical investigation and mathematical training (Wiesner-Hanks, 2019; Park, 2006). These exclusions were not incidental but constitutive of how coordination occurred.

Women like Hildegard of Bingen or Christine de Pizan developed alternative networks for knowledge production, but these remained marginalized relative to mainstream institutional coordination (Wiesner-Hanks, 2019). Similarly, much technical and translational work was performed by individuals of lower social status whose contributions enabled but rarely received credit for coordination achievements (Baxandall, 1972).

4.2 AQUINAS: SYSTEMATIC COORDINATION WITHIN THEOLOGICAL BOUNDARIES

Aquinas developed sophisticated methods for coordinating philosophical reasoning with theological commitment through explicit procedures that preserved the autonomy of different inquiry domains while establishing systematic relationships between them.

4.2.1 The Disputed Question Method as Coordination Procedure

The *quaestio disputata* format provided systematic procedures for handling apparent conflicts between different authorities and methods. Each question presents objections, cites contrary authorities, provides reasoned response, and addresses objections systematically. This format enabled coordination by making explicit the relationships between different types of evidence and argument.

However, this coordination operated within predetermined doctrinal boundaries. Where philosophical reasoning conflicted with revealed doctrine—Trinity, Incarnation, transubstantiation—theological authority ultimately prevailed. The method coordinated reason with faith but subordinated the former to the latter rather than establishing genuine parity.

The systematic character of Aquinas's approach provided transferable methodological insights about explicit argumentation, systematic objection-handling, and transparent reasoning processes that remain valuable for contemporary coordination efforts, independent of his specific theological commitments.

4.2.2 Analogical Predication as a Coordination Tool

Aquinas's doctrine of analogy attempted to solve fundamental coordination problems about how language and concepts could apply across radically different levels of reality—finite and infinite, material and spiritual, natural and supernatural. The theory of analogical predication enabled systematic coordination by providing principled ways to relate different domains without either reducing them to each other or treating them as completely unrelated (Stump, 2003; Davies, 2021).

The essence-existence distinction, adapted from Avicenna, provided tools for analyzing how particular things participate in universal structures while maintaining their distinctiveness (Davies, 2021). This enabled coordination between abstract conceptual analysis and concrete ontological commitment without reductive identification.

However, the analogical solution generated new problems about how analogical relationships work and what grounds their validity. Later philosophers, including Islamic critics like Averroes and modern analysts, argued that the analogical "middle" between univocal and equivocal predication remains unclear or incoherent.

4.2.3 Natural Law as Cross-Domain Coordination

Aquinas's natural law theory demonstrated systematic coordination across ethics, politics, jurisprudence, and theology through a hierarchical organization that preserved domain-specific methods while establishing systematic relationships. Eternal law, natural law, human law, and divine law formed coordinated levels rather than competing authorities.

This coordination enabled medieval canon law to integrate Roman jurisprudence with biblical mandates through philosophical reasoning, creating systematic approaches to practical decision-making that influenced later legal and political theory.

However, the hierarchical organization embedded specific value commitments about divine authority, natural teleology, and social hierarchy that contemporary democratic and pluralistic contexts cannot simply adopt. The coordination method requires separation from its specific hierarchical content.

4.3 LEONARDO: EMPIRICAL-AESTHETIC COORDINATION THROUGH VISUAL INVESTIGATION

Leonardo developed alternative coordination approaches that integrated observational investigation, technical problem-solving, and aesthetic representation through visual epistemology and biomimetic methodology (Kemp, 2019).

4.3.1 "Sapere Vedere" as Methodological Principle

Leonardo's principle of "knowing how to see" (*sapere vedere*) represented more than passive observation (Kemp, 2019; Capra, 2020). It involved trained perception guided by mathematical understanding, anatomical knowledge, and aesthetic sensitivity. This approach enabled coordination between empirical investigation and artistic representation without reducing either domain to the other.

The anatomical investigations, based on direct dissection of human cadavers, demonstrated systematic coordination of artistic representation with empirical investigation (O'Malley & Saunders, 1952; Kemp, 2019). The drawings provide both accurate scientific documentation and aesthetically compelling visual synthesis, enabling coordination between different validation criteria.

However, Leonardo's visual epistemology remained largely personal rather than institutional (Kemp, 2019). His investigations, confined to private notebooks, generated remarkable individual achievements but did not create transferable methodological protocols that others could systematically apply. The coordination remained at the level of individual practice rather than systematic method.

4.3.2 Biomimetic Investigation as Coordination Strategy

Leonardo's systematic study of natural forms as sources for technical innovation demonstrated productive coordination between biological investigation and engineering design. The studies of bird flight, water dynamics, and plant structures generated insights applicable to human technical problems while advancing understanding of natural phenomena (Vincent et al., 2006; Kemp, 2019).

This biomimetic approach enabled genuine cross-domain learning that avoided both reductive mechanism and vague organicism. Natural forms provided design principles that could be abstracted and applied technically while respecting the integrity of biological investigation.

However, Leonardo's biomimicry operated without an evolutionary understanding of why organisms exhibit particular features. This limited his ability to distinguish between essential design principles and historically contingent adaptations, reducing the systematic transferability of his approach.

4.3.3 Coordination Through Visual Modeling

Leonardo's coordination of mathematical proportion, empirical observation, and artistic representation through visual modeling provided coordination tools that enabled movement between abstract principles and concrete applications. The Vitruvian Man exemplifies this approach by coordinating classical proportion theory with anatomical observation through geometric construction (Kemp, 2019).

This visual coordination enabled insights unavailable through purely verbal or mathematical approaches while maintaining systematic relationships between different types of investigation. The drawings functioned as boundary objects that could be used across different domains without eliminating their distinctiveness.

However, Leonardo's visual coordination sometimes prioritized aesthetic coherence over empirical accuracy, limiting its reliability for systematic scientific investigation. The coordination worked better for heuristic discovery than for rigorous validation.

4.4 COMPARATIVE ANALYSIS: DIFFERENT COORDINATION STRATEGIES

Aquinas and Leonardo developed complementary but distinct coordination strategies that responded to different institutional contexts and problem sets.

4.4.1 Systematic vs. Exploratory Coordination

Aquinas pursued systematic coordination through explicit organizational frameworks that specified relationships between different domains and methods. This approach provided comprehensive coverage and systematic coherence while operating within clear institutional boundaries.

Leonardo pursued exploratory coordination through empirical investigation and visual modeling that remained open to unexpected discoveries and cross-domain connections. This approach enabled innovative insights and creative problem-solving while lacking systematic organizational frameworks.

Both approaches offer valuable coordination resources: systematic organization for comprehensive coverage and exploratory investigation for innovative discovery. Contemporary coordination efforts can benefit from both strategies while avoiding their respective limitations.

4.4.2 Hierarchical vs. Network Organization

Aquinas organized different domains hierarchically, with theology providing ultimate authority and other domains contributing within their proper spheres. This hierarchical coordination enabled systematic coordination while preserving domain-specific methods and expertise.

Leonardo worked with network-like connections between different domains, where insights from any area could inform others without predetermined priority relationships. This network approach enabled creative cross-fertilization while lacking systematic organizational principles.

Hierarchical coordination provides systematic organization but risks subordinating rather than genuinely integrating different domains. Network coordination enables creative connections but may lack systematic coherence. Contemporary approaches can combine both organizational strategies while avoiding their respective limitations.

4.5 ACHIEVEMENTS WITHIN CONSTRAINTS

Both figures achieved genuine coordination innovations within their historical contexts while operating under significant constraints that limited the scope and transferability of their achievements.

4.5.1 Methodological Innovations

Aquinas developed transferable procedures for systematic argumentation, explicit objection-handling, and analogical reasoning across domains. These methodological innovations continue to inform systematic philosophical investigation and interdisciplinary coordination efforts.

Leonardo developed innovative approaches to visual investigation, biomimetic design, and empirical-aesthetic coordination that anticipated later scientific methodology and continue to influence contemporary design thinking and biomimicry.

However, both sets of innovations operated within institutional and cultural constraints that shaped their development and limited their transferability to different contexts. The methodological insights require careful extraction from their original contexts and adaptation to contemporary coordination challenges.

4.5.2 Institutional and Cultural Limitations

Aquinas's coordination operated within ecclesial orthodoxy that predetermined acceptable conclusions and limited the autonomy of philosophical reasoning. This constraint enabled his systematic approach while preventing genuine open-ended investigation.

Leonardo's coordination operated within patronage relationships that provided resources while constraining research directions toward practical and spectacular applications rather than systematic theoretical development.

Both sets of constraints enabled certain coordination achievements while preventing others. Contemporary applications must acknowledge these scope limitations and adapt the methodological insights to different institutional contexts.

4.5.3 Systematic Exclusions

The systematic exclusion of women, non-Christians, and lower-status individuals from full participation shaped the forms of coordination that emerged rather than simply limiting their scope. Apparent coordination often reflected restricted participation rather than genuine synthesis of diverse perspectives.

These exclusions were constitutive of how coordination occurred rather than incidental features that can be ignored. Contemporary appropriations must address rather than reproduce exclusionary patterns while adapting coordination methods to more inclusive institutional contexts.

4.6 CONTEMPORARY RELEVANCE: BOUNDED LESSONS

Despite historical limitations, both figures offer methodological insights relevant for contemporary coordination challenges when properly contextualized and adapted.

4.6.1 Procedural Coordination

Aquinas's systematic procedures for handling conflicting evidence, explicit argumentation, and analogical reasoning provide resources for contemporary interdisciplinary coordination. The disputed question format offers structured approaches to managing disagreement while maintaining systematic relationships between different positions.

However, these procedures require adaptation to contemporary democratic and pluralistic contexts that cannot simply adopt hierarchical authority structures. The procedural insights must be separated from specific theological and metaphysical commitments.

4.6.2 Visual and Embodied Coordination

Leonardo's coordination of visual investigation, embodied experience, and aesthetic judgment provides resources for contemporary approaches to scientific visualization, design thinking, and biomimicry. His emphasis on seeing, making, and iterative investigation anticipates contemporary recognition of embodied and situated cognition.

However, Leonardo's individual approach requires institutional development to become systematically transferable. Contemporary applications must develop systematic protocols and validation criteria that preserve his insights while enabling collaborative investigation.

4.6.3 Boundary Object Development

Both figures created artifacts—systematic treatises, illustrated notebooks—that could function across different domains while preserving their distinctiveness. These boundary objects enabled coordination without forcing theoretical unification.

Contemporary coordination efforts can learn from their approaches to developing shared resources that enable communication across different domains while respecting methodological differences. However, these boundary objects require democratic development rather than individual creation.

4.7 Critical Assessment: Partial Achievements and Persistent Problems

A realistic assessment must acknowledge both genuine coordination innovations and significant limitations that constrain contemporary applicability.

4.7.1 Incomplete Synthesis

Neither figure achieved the comprehensive coordination often attributed to them. Aquinas's coordination subordinated philosophy to theology rather than establishing genuine parity. Leonardo's coordination remained largely personal rather than systematic and transferable.

The apparent "synthesis" in both cases masked unresolved tensions and institutional compromises rather than achieving genuine theoretical coordination. Contemporary coordination efforts should expect similar limitations and focus on productive partial coordination rather than comprehensive theoretical unification.

4.7.2 Institutional Dependence

Both coordination approaches depended heavily on particular institutional contexts—university theology, court patronage—that provided resources while imposing constraints. The apparent success of their coordination reflected institutional possibilities as much as individual methodological innovation.

Contemporary coordination efforts must similarly work within institutional constraints while attempting to modify institutional structures to enable more productive coordination. The historical examples illuminate both possibilities and limitations of coordination within institutional contexts.

4.7.3 Cultural Specificity

Both approaches reflected specific cultural assumptions about knowledge, authority, and social organization that limit their transferability across cultural contexts. Medieval Christian assumptions about divine authority and Renaissance assumptions about individual genius shaped their coordination possibilities.

Contemporary global coordination efforts must address cultural differences more systematically than either historical approach achieved.

The methodological insights require careful adaptation to different cultural contexts rather than direct transfer.

4.8 METHODOLOGICAL PRINCIPLES FOR CONTEMPORARY APPLICATION

Despite historical limitations, several methodological principles emerge as potentially transferable to contemporary coordination challenges:

Explicit Coordination Procedures: Develop systematic procedures for handling disagreement and conflict between different domains rather than assuming spontaneous harmony.

Analogical and Visual Bridging: Create explicit bridge concepts and visual tools that enable movement between different domains while preserving their distinctiveness.

Hierarchical Organization with Democratic Modification: Use systematic organizational principles while avoiding authoritarian subordination of some domains to others.

Boundary Object Development: Create shared resources that different domains can use without eliminating their methodological differences.

Iterative Investigation: Combine systematic organization with exploratory investigation that remains open to unexpected discoveries and connections.

Inclusion Accountability: Systematically address exclusions that may undermine coordination by restricting participation and perspective diversity.

Institutional Adaptation: Modify institutional structures to enable coordination while working within existing institutional constraints.

4.9 CONCLUSION: HISTORICAL EXPERIMENTS IN CONTEMPORARY CONTEXT

Aquinas and Leonardo represent sophisticated experiments in coordination across different domains of inquiry within particular historical constraints. Their achievements—systematic argumentation procedures, analogical reasoning tools, visual investigation methods, biomimetic design principles—offer methodological resources for contemporary coordination challenges.

However, these achievements occurred within institutional and cultural constraints that shaped their possibilities and limited their transferability.

Contemporary applications must extract methodological insights while addressing rather than reproducing the exclusionary and hierarchical aspects of medieval and Renaissance approaches.

The goal should be coordination methods that serve democratic and inclusive values rather than elite knowledge production within hierarchical social structures. This requires systematic institutional and cultural adaptation rather than direct appropriation of historical methods.

Both figures demonstrate that productive coordination across domains requires creative methodological innovation, institutional support, and systematic attention to scope limitations. Their experiments provide resources for contemporary coordination efforts while illustrating persistent challenges that continue to shape integrative work.

CHAPTER 5: AMBITIOUS COORDINATION PROJECTS - LEIBNIZ AND GOETHE'S SYSTEMATIC EXPERIMENTS

ABSTRACT

This chapter examines Gottfried Wilhelm Leibniz (1646-1716) and Johann Wolfgang von Goethe (1749-1832) as historically significant experiments in systematic knowledge coordination whose ambitious projects achieved partial success while revealing persistent limitations of comprehensive synthesis. Leibniz's characteristica universalis and monadological metaphysics attempted formal-logical unification across domains, while Goethe's morphological method and phenomenological empiricism pursued coordination through disciplined observation of natural forms (Antognazza, 2009; Garber, 2009). Rather than treating these as successful universal models, we analyze their specific methodological innovations, institutional constraints, and scope limitations. Both projects generated transferable coordination principles—systematic symbolization, analogical reasoning, phenomenological observation, processual thinking—while failing to achieve their comprehensive synthetic ambitions (Antognazza, 2009; Richards, 2002). The analysis extracts bounded lessons for contemporary coordination challenges while avoiding anachronistic attribution and acknowledging the productive specialization that their failed unifications helped motivate.

5.1 HISTORICAL CONTEXT: BETWEEN MECHANISM AND ORGANISM

The seventeenth through nineteenth centuries witnessed competing responses to the perceived fragmentation of knowledge as natural philosophy divided into specialized sciences. Leibniz and Goethe represent ambitious attempts to preserve systematic unity while accommodating increasing empirical complexity and disciplinary differentiation.

Leibniz developed his systematic philosophy during the period when Cartesian mechanism and Newtonian physics were establishing mathematical approaches as the standard for natural knowledge. His coordination project attempted to preserve the comprehensiveness of traditional metaphysics while incorporating insights from the new mathematical sciences. However, this project operated within significant constraints—theological commitments, limited empirical knowledge, pre-modern institutional contexts—that shaped both its achievements and limitations.

Goethe's scientific work emerged during the Romantic reaction against Enlightenment mechanism, yet he explicitly rejected both reductive mechanism and anti-scientific romanticism (Richards, 2002). His morphological investigations sought systematic principles governing natural transformation without reducing life to mechanical processes. However, his approach developed largely outside established scientific institutions and achieved limited acceptance among contemporary practitioners (Richards, 2002).

The temporal separation between these figures (over a century) and their different cultural contexts complicate claims about methodological continuity or complementarity. Their coordination projects responded to different intellectual challenges within different institutional constraints rather than representing a unified tradition.

5.2 LEIBNIZ: FORMAL COORDINATION AND ITS LIMITS

Leibniz pursued coordination through two interconnected projects: developing a universal symbolic language (characteristica universalis) and constructing a systematic metaphysics (monadology) that could ground unified knowledge in fundamental principles (Antognazza, 2009; Garber, 2009; Leibniz, 1989).

5.2.1 The Characteristica Universalis Project

Leibniz envisioned a universal symbolic language capable of expressing all possible thoughts with sufficient precision that intellectual disputes could be resolved through calculation rather than argumentation. This project aimed to create formal coordination across all domains of knowledge through systematic symbolization and logical inference (Antognazza, 2009; Garber, 2009).

The project achieved limited success in specific domains—binary arithmetic, symbolic logic, differential calculus—while failing as a comprehensive universal language (Antognazza, 2009). Leibniz's innovations in mathematical notation and logical symbolism proved durable and influential, but the broader project of reducing all reasoning to calculation proved impossible to complete.

Contemporary formal logic and computer science developed through different routes that owe little direct debt to Leibnizian programs. While Leibniz's insights about symbolic representation and mechanical calculation proved prescient, connecting his work directly to modern computational theory involves significant anachronistic overreach. The genuine contribution lies in establishing systematic symbolization as a coordination tool rather than in specific anticipations of later developments.

5.2.2 Monadological Metaphysics

The monadology attempted to provide metaphysical foundations for systematic knowledge by grounding all phenomena in simple substances (monads) coordinated through pre-established harmony (Leibniz, 1989; Garber, 2009; Russell, 1937). This framework sought to preserve both mechanistic regularity and qualitative experience within a comprehensive systematic account.

The monadological framework addressed genuine coordination challenges—how to relate different levels of organization, how to preserve both causal regularity and phenomenological richness, how to account for unity within diversity. Leibniz's solutions involved sophisticated conceptual innovations including the principle of sufficient reason, pre-established harmony, and graduated perfection that influenced subsequent philosophical development (Leibniz, 1989; Russell, 1937; Garber, 2009).

However, the system depended on theological assumptions about divine coordination that many contemporary naturalists cannot accept. The pre-established harmony solution to mind-body coordination created more philosophical problems than it solved, while the principle of sufficient reason led to deterministic consequences that conflicted with genuine novelty and freedom (Leibniz, 1989; Russell, 1937; Garber, 2009).

5.2.3 Methodological Contributions and Limitations

Leibniz's genuine methodological contributions include systematic attention to logical consistency, development of analogical reasoning principles, and recognition that different domains require coordinated but distinct explanatory approaches (Antognazza, 2009). His emphasis on formal precision and systematic organization provided transferable tools for coordination across domains.

However, the systematic character of Leibniz's coordination came at the cost of empirical testability and institutional viability. His metaphysical commitments prevented genuine engagement with emerging empirical sciences, while his systematic ambitions outran what logical and mathematical tools could deliver. The project's failure illuminates the difficulty of achieving comprehensive theoretical unification.

5.3 GOETHE: PHENOMENOLOGICAL COORDINATION AND ITS BOUNDARIES

Goethe developed alternative coordination approaches through morphological observation and phenomenological empiricism that sought systematic principles through disciplined engagement with natural phenomena rather than formal theorization.

5.3.1 Morphological Method

Goethe's morphological investigations of plant and animal forms attempted to discover systematic transformation principles through sustained observation of development and metamorphosis. This approach sought coordination between different organic forms through identification of underlying patterns (Urphänomene) manifest in variation (Richards, 2002).

The morphological method achieved genuine insights about organic unity and transformation that influenced subsequent biological development. Goethe's recognition of plant organs as modifications of fundamental forms, his attention to developmental constraint and morphological coordination, and his systematic comparison of related forms provided methodological resources that proved valuable for later biological research.

However, the method remained largely qualitative and dependent on individual observational skill rather than providing systematic protocols that others could replicate. Goethe's morphological insights, while valuable, did not provide predictive frameworks or quantitative methods that could compete with mathematical approaches to biological investigation.

5.3.2 Phenomenological Empiricism

Goethe's "delicate empiricism" attempted to integrate rigorous observation with aesthetic perception through disciplined phenomenological investigation (Seamon & Zajonc, 1998). This approach recognized qualitative dimensions of experience that mechanical observation alone could not capture while maintaining systematic attention to natural phenomena.

The phenomenological approach developed systematic protocols for trained observation that remain relevant for certain types of investigation, particularly in ecological and organismal biology where pattern recognition and qualitative assessment prove valuable. Goethe's emphasis on observer development and participatory engagement with phenomena anticipated later phenomenological and ecological approaches.

However, the method's dependence on cultivated individual judgment and its weak reproducibility criteria limited its acceptance within scientific communities increasingly oriented toward quantitative measurement and experimental control. The approach worked better for exploratory investigation than for hypothesis testing or predictive modeling.

5.3.3 Color Theory and Scientific Controversy

Goethe's Farbenlehre attempted to integrate physical optics with phenomenological investigation of color experience, challenging Newton's mathematical approach through systematic attention to qualitative color phenomena (Westfall, 1980; Sepper, 1988). This project exemplified his broader methodological program of coordinating scientific investigation with experiential richness.

While Goethe's color theory achieved insights about color perception and phenomenology that remain relevant for vision science and aesthetic theory, his rejection of mathematical optics proved scientifically counterproductive. Newton's approach prevailed because it delivered quantitative predictions and experimental control that Goethe's phenomenological alternative could not match.

The color theory controversy illuminates the difficulty of integrating phenomenological and mathematical approaches when they yield incompatible conclusions. Goethe's insights about color experience require coordination with rather than replacement of mathematical optics, but his systematic ambitions prevented such coordination.

5.4 COMPARATIVE ASSESSMENT: DIFFERENT ROUTES TO FAILED SYNTHESIS

Despite their different methodological emphases, both Leibniz and Goethe pursued comprehensive systematic coordination that ultimately proved unsustainable within their institutional and intellectual contexts.

5.4.1 Systematic Ambitions and Scope Limitations

Both figures aimed for comprehensive coordination across all domains of knowledge rather than more modest coordination within specific problem areas. Leibniz sought universal principles applicable across mathematics, physics, metaphysics, theology, and law. Goethe pursued systematic coordination across art, science, philosophy, and practical life.

These comprehensive ambitions enabled certain innovative insights by forcing attention to connections across traditionally separated domains. However, the systematic character of their projects also created unrealistic demands for theoretical unification that outran available conceptual and empirical resources.

The failure of their systematic ambitions does not invalidate their specific methodological contributions, but it does suggest the importance of more modest coordination goals that preserve domain-specific expertise while enabling productive boundary-crossing work.

5.4.2 Institutional and Cultural Constraints

Both projects developed within institutional and cultural contexts that provided certain resources while imposing significant constraints. Leibniz operated within court patronage systems and theological frameworks that enabled his systematic work while constraining its empirical engagement. Goethe worked largely outside established scientific institutions, which provided freedom for innovative investigation while limiting systematic uptake of his insights (Richards, 2002).

The institutional marginality of both projects suggests the difficulty of achieving systematic coordination within existing disciplinary structures. However, this marginality also limited their ability to influence ongoing scientific development or establish sustainable coordination practices.

Contemporary coordination efforts must work within existing institutional constraints while attempting to modify institutional structures to enable more productive boundary-crossing work. The historical examples illustrate both the importance and the difficulty of institutional innovation for coordination success.

5.4.3 Productive Failure and Subsequent Development

The failure of these comprehensive coordination projects contributed indirectly to subsequent intellectual development by clarifying the scope limitations of systematic unification and motivating more modest approaches to cross-domain coordination.

Leibniz's logical and mathematical innovations proved valuable for subsequent development even though his comprehensive metaphysical system was abandoned. Goethe's morphological insights influenced biological research even though his anti-mathematical stance was rejected. The productive aspects of their work could be preserved while their systematic overreach was corrected.

This pattern suggests that ambitious coordination projects may prove valuable even when they fail to achieve their systematic goals, provided their specific methodological contributions can be extracted and adapted to more modest coordination challenges.

5.5 CONTEMPORARY RELEVANCE: BOUNDED LESSONS

Despite their historical limitations, both figures offer methodological insights relevant for contemporary coordination challenges when properly contextualized and adapted.

5.5.1 Formal Coordination Tools

Leibniz's work on symbolic representation, logical consistency, and analogical reasoning provides resources for contemporary approaches to knowledge representation, formal ontologies, and logical coordination across domains (Antognazza, 2009). His insights about the importance of precise symbolization and systematic inference remain relevant for formal approaches to interdisciplinary coordination.

However, contemporary applications must avoid the metaphysical overreach that limited Leibniz's project while adapting his formal insights to different institutional and technological contexts. The goal should be coordination tools that enable communication across domains without imposing comprehensive theoretical unification.

5.5.2 Phenomenological Investigation

Goethe's morphological method and phenomenological empiricism provide resources for approaches to systematic observation, pattern recognition, and qualitative investigation that complement quantitative methods without replacing them. His insights about observer development and participatory investigation remain relevant for certain types of research.

However, contemporary applications must address the reproducibility and validation challenges that limited Goethe's scientific acceptance while preserving the insights about qualitative investigation that his approach generated. The goal should be systematic protocols for qualitative investigation that meet contemporary scientific standards.

5.5.3 Coordination Design Principles

Both figures developed systematic approaches to coordination that provide transferable design principles for contemporary coordination efforts:

Explicit Symbolization: Develop shared representational systems that clarify assumptions and enable systematic inference across domains.

Analogical Reasoning: Use systematic analogical thinking to identify patterns and relationships across different domains while respecting their distinctiveness (Antognazza, 2009).

Phenomenological Attention: Develop systematic protocols for qualitative observation and pattern recognition that complement quantitative methods.

Processual Thinking: Emphasize transformation and development rather than static categories when attempting to coordinate across different domains (Richards, 2002).

Systematic Organization: Provide explicit organizational frameworks for handling complexity without eliminating relevant differences.

5.6 CRITICAL ASSESSMENT: WHY SYSTEMATIC COORDINATION FAILED

A realistic assessment must acknowledge the specific reasons why these ambitious coordination projects failed to achieve their systematic goals despite their genuine methodological innovations.

5.6.1 Institutional Selection Pressures

Scientific institutions increasingly favored approaches that provided quantitative prediction, experimental control, and reproducible results—features that neither Leibniz's metaphysical system nor Goethe's phenomenological method could consistently deliver (Westfall, 1980). The institutional development of modern science selected for specialized approaches that proved more reliable for specific investigative purposes.

This institutional selection was not arbitrary but reflected the superior performance of specialized mathematical and experimental approaches for most scientific purposes. The coordination projects failed partly because they could not compete with more focused disciplinary approaches in terms of practical results.

Contemporary coordination efforts must acknowledge these institutional selection pressures and develop coordination approaches that enhance rather than compete with specialized disciplinary expertise. The goal should be coordination that enables rather than replaces disciplinary excellence.

5.6.2 Theoretical Overreach

Both projects attempted to ground their coordination efforts in comprehensive theoretical systems that made claims beyond what available evidence could support. Leibniz's pre-established harmony and Goethe's morphological archetypes involved theoretical commitments that could not be empirically validated or systematically applied (Leibniz, 1989; Russell, 1937).

This theoretical overreach created unnecessary obstacles to coordination by requiring acceptance of contested metaphysical assumptions rather than focusing on practical coordination tools that could work across different theoretical frameworks.

Contemporary coordination efforts should focus on methodological coordination rather than theoretical unification, developing tools and protocols that can work across different theoretical commitments rather than requiring comprehensive systematic agreement.

5.6.3 Individual vs. Institutional Approaches

Both projects remained largely individual achievements rather than establishing sustainable institutional practices for coordination. Leibniz's systematic philosophy and Goethe's morphological method depended heavily on individual genius rather than creating transferable protocols that others could systematically apply.

This limitation meant that their coordination achievements could not be systematically transmitted or institutionally developed, leading to their marginalization as disciplinary specialization advanced. The failure to create sustainable institutional practices for coordination contributed to the projects' ultimate failure.

Contemporary coordination efforts must focus on developing institutional practices and collaborative protocols rather than depending on individual systematic vision. The goal should be sustainable coordination practices that can persist across different individuals and institutional contexts.

5.7 METHODOLOGICAL LESSONS FOR CONTEMPORARY COORDINATION

Despite their ultimate failure as comprehensive coordination projects, both figures developed methodological insights that remain relevant for more modest contemporary coordination challenges.

5.7.1 Coordination Without Unification

Both projects demonstrate the difficulty of achieving comprehensive theoretical unification while also showing the value of more modest coordination approaches. The transferable insights involve coordination tools and protocols rather than systematic theoretical frameworks.

Contemporary coordination efforts should aim for productive working relationships between different approaches rather than comprehensive theoretical synthesis. This requires developing explicit protocols for communication, translation, and collaboration across domains without eliminating their methodological differences.

5.7.2 Formal and Phenomenological Coordination

The different emphases of Leibniz (formal-logical) and Goethe (phenomenological-aesthetic) suggest that productive coordination may require both types of approaches working in complementary ways rather than seeking synthesis within a single framework.

Contemporary coordination efforts can benefit from both formal tools for precision and systematicity and phenomenological approaches for qualitative richness and pattern recognition. The challenge is coordinating these different approaches rather than choosing between them or forcing their synthesis.

5.7.3 Scope Limitation and Domain Respect

Both projects suffered from insufficient attention to scope limitations and domain boundaries. Their systematic ambitions prevented appropriate recognition of where their coordination methods worked well and where they encountered legitimate obstacles.

Contemporary coordination efforts should begin with explicit scope analysis, identifying where coordination is beneficial and feasible while respecting domains where specialized approaches may be more appropriate. This requires replacing comprehensive systematic ambitions with more targeted coordination goals.

5.8 HISTORICAL SIGNIFICANCE AND CONTEMPORARY APPLICATION

Leibniz and Goethe remain historically significant not as successful models of systematic coordination but as instructive experiments whose achievements and failures illuminate both the possibilities and limitations of coordination across domains.

5.8.1 Productive Specialization

The failure of their comprehensive coordination projects contributed to the productive development of disciplinary specialization that enabled the remarkable scientific and technological achievements of subsequent centuries. Their failures helped clarify the value of focused expertise and the difficulty of comprehensive synthesis.

This suggests that coordination efforts should aim to enhance rather than replace disciplinary specialization, developing tools and protocols that enable productive collaboration between specialists rather than attempting to transcend specialization through comprehensive coordination.

5.8.2 Persistent Coordination Challenges

The coordination challenges that Leibniz and Goethe addressed—relating different levels of organization, coordinating quantitative and qualitative approaches, integrating formal and experiential dimensions—remain relevant for contemporary interdisciplinary work.

Their methodological innovations provide resources for addressing these persistent challenges while their failures illustrate the importance of modest goals, explicit scope limitations, and institutional sustainability for coordination success.

5.8.3 Coordination as Ongoing Process

Both figures demonstrate that coordination is better understood as an ongoing process of developing coordination tools and practices rather than as achievement of final systematic synthesis. Their projects succeeded in generating transferable methodological insights while failing to achieve comprehensive theoretical unification.

This suggests that contemporary coordination efforts should focus on developing sustainable practices for ongoing collaboration rather than seeking final synthetic solutions to interdisciplinary challenges. The goal should be continuous improvement of coordination capabilities rather than comprehensive coordination.

5.9 CONCLUSION: LEARNING FROM AMBITIOUS FAILURE

Leibniz and Goethe represent historically significant experiments in systematic knowledge coordination whose ambitious goals, methodological innovations, and ultimate limitations provide important lessons for contemporary coordination challenges.

Their achievements—systematic symbolization, analogical reasoning, morphological observation, phenomenological investigation—offer transferable tools for coordination across domains (Antognazza, 2009). Their failures—theoretical overreach, institutional marginality, insufficient scope limitation—illustrate persistent obstacles to comprehensive coordination that contemporary efforts must acknowledge and address.

The historical analysis suggests that productive coordination requires modest goals, explicit scope limitations, institutional sustainability, and respect for disciplinary expertise rather than comprehensive systematic synthesis. The methodological insights of these ambitious coordination projects remain valuable when adapted to more realistic coordination goals within contemporary institutional contexts.

Rather than seeking to emulate their systematic ambitions, contemporary coordination efforts should learn from both their innovations and their limitations, developing coordination approaches that enhance rather than replace specialized expertise while enabling productive collaboration across disciplinary boundaries where such collaboration proves beneficial.

CHAPTER 6: TEILHARD DE CHARDIN AND MUHAMMAD IQBAL - EVOLUTIONARY COORDINATION AND ITS DISCONTENTS: A CRITICAL ASSESSMENT

ABSTRACT

Pierre Teilhard de Chardin (1881-1955) and Muhammad Iqbal (1877-1938) represent ambitious attempts to reconcile evolutionary science with religious thought during a pivotal moment in intellectual history (Teilhard de Chardin, 2008; Iqbal, 2013). Their projects emerged when the implications of Darwinian evolution were still reverberating through religious communities, and both sought to demonstrate that evolutionary theory could enhance rather than undermine theological understanding (Dennett, 1995; Gould, 1999). However, the scholarly reception of their work reveals fundamental tensions between their synthetic ambitions and the methodological requirements of both rigorous science and systematic theology.

This chapter examines their contributions through three analytical lenses: their historical context and methodological innovations, the substantive claims they advanced, and their contemporary relevance. While acknowledging the genuine insights embedded within their frameworks, this analysis identifies critical weaknesses that limit their viability as comprehensive syntheses. The central argument developed here is that while both thinkers made valuable contributions to the broader project of science-religion dialogue, their specific attempts to ground teleological claims in evolutionary science remain fundamentally problematic.

6.1 HISTORICAL CONTEXT AND METHODOLOGICAL CONTRIBUTIONS

6.1.1 The Crisis of Coordination

The early twentieth century presented religious thinkers with an unprecedented challenge: how to maintain meaningful theological discourse in light of scientific developments that appeared to render traditional cosmologies obsolete. The mechanistic worldview emerging from modern physics and biology seemed to leave no room for purpose, meaning, or divine action within natural processes (Dennett, 1995).

Teilhard approached this challenge from his dual expertise as a Jesuit priest and trained paleontologist. His direct participation in significant fossil discoveries, including work on Peking Man, provided him with scientific credentials that distinguished his theological speculations from purely armchair philosophy (Aczel, 2007). Iqbal brought comparable intellectual authority through his philosophical training at Cambridge and Munich, combined with deep grounding in Islamic jurisprudence and mystical traditions (Iqbal, 2013; Rahman, 1982).

6.1.2 Methodological Innovations

Both thinkers developed sophisticated epistemological frameworks that integrated multiple sources of knowledge. Teilhard's "phenomenological naturalism" attempted to combine empirical observation with first-person conscious experience and contemplative insight (Thompson, 2007; Haught, 2017). This triadic methodology refused both reductive materialism's exclusion of consciousness and traditional dualism's separation of matter and spirit.

Iqbal's approach, which can be termed "pragmatic-Islamic synthesis," similarly integrated reason ('aql), empirical experience (tajriba), and intuitive insight (wujdan) (Iqbal, 2013; Koshul, 2014). His methodology demonstrated how Islamic intellectual resources could engage constructively with Western philosophy and modern science without abandoning distinctive theological commitments.

These methodological contributions remain significant. Both thinkers recognized that adequate responses to scientific challenges required more than defensive apologetics; they demanded new integrative approaches capable of honoring the integrity of different domains while identifying meaningful connections between them.

6.2 SUBSTANTIVE CLAIMS AND CRITICAL ASSESSMENT

6.2.1 Teilhard's Complexity-Consciousness Hypothesis

Teilhard's central claim concerned a fundamental correlation between material complexity and consciousness throughout evolutionary history. His "law of complexity-consciousness" proposed that increasing neural complexity corresponds systematically to enhanced conscious experience, culminating in human reflective awareness and pointing toward future developments in collective consciousness (Teilhard de Chardin, 2008; Deane-Drummond, 2022).

Strengths: This hypothesis anticipated contemporary discussions in consciousness studies, particularly Integrated Information Theory's correlation between information coordination and consciousness (Tononi & Koch, 2015). Teilhard's recognition that consciousness poses genuine explanatory challenges for reductive materialism remains relevant to current debates in philosophy of mind.

Critical Problems: The complexity-consciousness correlation faces several decisive objections:

Operational Definition Failures: Neither "complexity" nor "consciousness" receives sufficiently precise definition to enable empirical testing (Adami, 2016; Tononi & Koch, 2015). Different measures of complexity (structural, computational, thermodynamic) yield inconsistent predictions.

Counterexamples: Numerous evolutionary lineages exhibit decreased complexity while maintaining sophisticated behaviors. Parasites often evolve simpler forms; colonial organisms can be highly successful with reduced individual complexity (Grosberg & Strathmann, 2007).

Anthropocentric Bias: The correlation appears to hold only when complexity is defined in ways that privilege traits leading toward human cognition, revealing circular reasoning rather than objective natural law (Gould, 1996).

6.2.2 Iqbal's Evolutionary Theism

Iqbal's reconstruction of Islamic thought proposed that divine creative action operates through evolutionary processes rather than despite them. His concept of khudi (dynamic selfhood) suggested that consciousness at all levels participates in cosmic creativity, making evolution the temporal manifestation of divine purpose (Iqbal, 2013; Koshul, 2014).

Strengths: Iqbal's process theology anticipated developments in contemporary philosophy of religion while maintaining continuity with Islamic theological traditions. His emphasis on temporal becoming and emergent novelty provides resources for understanding divine action in naturalistic terms.

Critical Problems:

Theological Inconsistencies: Iqbal's evolutionary framework conflicts with traditional Islamic doctrines of divine transcendence and complete knowledge (Nasr, 1989; Rahman, 1982). If God requires temporal process to achieve divine purposes, this suggests limitations inconsistent with classical theism.

Scriptural Tensions: Evolutionary interpretations of creation conflict with literal readings of Quranic texts accepted by mainstream Islamic scholarship, limiting Iqbal's influence within traditional communities (Nasr, 1989; Rahman, 1982).

Causal Overdetermination: Iqbal never adequately explains how divine creative action and natural selection can both be causally efficacious without conflict (Dennett, 1995).

6.2.3 The Orthogenesis Problem

Both thinkers commit what can be termed the "orthogenesis fallacy"—the assumption that evolution exhibits inherent directionality toward greater complexity, consciousness, or perfection (Gould, 1996; Simpson, 1944). This assumption conflicts fundamentally with evolutionary biology's core insights:

Scientific Objections:

Natural selection operates locally and contingently, not globally and directionally (Futuyma & Kirkpatrick, 2017)

Evolutionary outcomes depend on environmental pressures that change unpredictably

No mechanism exists for evolution to "aim" toward future states

Successful organisms often evolve toward greater simplicity rather than complexity (McCutcheon & Moran, 2012)

Philosophical Objections:

Directional interpretations conflate retrospective pattern-recognition with prospective purpose

Anthropocentric bias distorts evaluation of evolutionary "success"

Teleological language illegitimately imports intentionality into natural processes (Dennett, 1995)

6.3 CONTEMPORARY RELEVANCE AND REFORMULATION

6.3.1 Salvageable Insights

Despite fundamental problems with their core claims, both thinkers contributed insights that remain relevant when properly reformulated:

Emergent Complexity: While rejecting law-like directionality, evolutionary history does exhibit statistical tendencies toward increased maximum complexity in some lineages (Gould, 1996; McShea & Brandon, 2010). This can be explained through neutral evolution and niche-construction without invoking teleology.

Consciousness as Natural Phenomenon: Both thinkers correctly identified consciousness as a natural phenomenon requiring explanation rather than elimination (Tononi & Koch, 2015; Thompson, 2007). Contemporary neuroscience and philosophy of mind continues grappling with problems they recognized.

Meaning-Making Processes: Their emphasis on meaning-making as a fundamental feature of conscious systems anticipates enactive approaches to cognition and cultural evolution studies (Thompson, 2007; Varela et al., 1991).

Integrative Methodology: Their triadic epistemologies provide models for interdisciplinary research that avoids both reductive scientism and anti-empirical spiritualism (Mitchell, 2009; Deacon, 2011).

6.3.2 Reformulation Strategies

Several contemporary research programs build on their insights while avoiding teleological commitments:

Information-Theoretic Biology: Studies of information flow in biological systems can explain apparent directionality without invoking purpose or foresight (Adami, 2016).

Cultural Evolution: Mechanisms of cultural transmission and selection explain noospheric developments without requiring biological orthogenesis (Boyd & Richerson, 1985; Mesoudi, 2011).

Systems Theory: Emergence and self-organization in complex systems provide naturalistic explanations for apparent teleological phenomena (Mitchell, 2009; Deacon, 2011).

Process Philosophy: Contemporary process thought maintains their emphasis on temporal becoming while rejecting predetermined endpoints (Haught, 2017).

6.4 RECEPTION HISTORY AND ONGOING DEBATES

6.4.1 Scientific Community Response

The scientific community has largely rejected teleological interpretations of evolution while incorporating valuable empirical contributions (Futuyma & Kirkpatrick, 2017; Gould, 1996). Teilhard's paleontological work remains scientifically respected even as his theoretical framework is dismissed. This pattern suggests the importance of distinguishing between observational insights and theoretical interpretations.

6.4.2 Religious Community Reception

Both thinkers faced significant resistance from religious orthodoxies. The Catholic Church censored Teilhard's works during his lifetime, while traditional Islamic scholars rejected Iqbal's evolutionary framework (Aczel, 2007; Haught, 2021; Rahman, 1982; Nasr, 1989). However, both have influenced progressive theological movements that seek accommodation with scientific worldviews.

6.4.3 Philosophical Assessment

Contemporary philosophy of science largely supports the critical assessment presented here. The consensus recognizes their historical importance while rejecting their specific claims about evolutionary directionality. However, their methodological innovations continue influencing interdisciplinary approaches to consciousness studies and science-religion dialogue.

6.5 CONCLUSION: QUALIFIED ASSESSMENT

Teilhard de Chardin and Muhammad Iqbal made significant historical contributions to science-religion dialogue by demonstrating that evolutionary theory need not be perceived as inherently antireligious. Their methodological innovations in integrating multiple sources of knowledge remain valuable for contemporary interdisciplinary research. However, their specific attempts to ground teleological claims in evolutionary science must

be judged unsuccessful (Futuyma & Kirkpatrick, 2017; Dennett, 1995).
The most defensible approach to their legacy involves:

Retaining their insights about consciousness as a genuine natural phenomenon requiring explanation

Reformulating their complexity-consciousness correlations as statistical tendencies rather than universal laws

Rejecting their claims about evolutionary directionality and orthogenesis

Preserving their methodological commitments to integrative epistemology

Translating their theological insights into frameworks compatible with contemporary scientific understanding

This qualified assessment neither dismisses their contributions entirely nor accepts them uncritically. Instead, it recognizes both their genuine insights and serious limitations while identifying resources for contemporary research programs that maintain scientific rigor without eliminating meaning, consciousness, or value from our understanding of nature.

The ultimate significance of Teilhard and Iqbal may lie not in the specific syntheses they proposed, but in their demonstration that the relationship between science and religion admits of creative possibilities beyond simple conflict or compartmentalization. Their failures remain instructive, and their successes continue inspiring new generations of scholars committed to integrative understanding of human existence within cosmic evolution.

CHAPTER 7: JUNG AND BOHM - THE LIMITS AND POSSIBILITIES OF INTERDISCIPLINARY COORDINATION

ABSTRACT

The relationship between Carl Gustav Jung's analytical psychology and David Bohm's theoretical physics presents a compelling case study in the challenges of interdisciplinary synthesis. Both figures developed sophisticated approaches to understanding consciousness, wholeness, and the participatory nature of knowledge that appear to offer remarkable convergences across disciplinary boundaries. However, a critical examination reveals that their apparent methodological alignment may mask fundamental differences in validation criteria, evidence standards, and conceptual precision that limit the viability of their coordination.

This chapter argues that while Jung and Bohm made significant contributions within their respective domains, attempts to synthesize their work into a unified theory of "participatory consciousness" face insurmountable methodological problems. Their convergences often reflect superficial linguistic similarities rather than substantive theoretical alignment, and their most ambitious claims venture beyond what either psychological observation or physical theory can empirically support. Nevertheless, a careful analysis of their contributions identifies valuable methodological insights that remain relevant for contemporary interdisciplinary research when properly circumscribed.

7.1 JUNG'S ANALYTICAL PSYCHOLOGY: ACHIEVEMENTS AND LIMITATIONS

7.1.1 Clinical Method and Empirical Foundations

Jung's most defensible contributions emerge from his systematic clinical observations and therapeutic innovations rather than his broader theoretical speculations. His development of psychological typology, dream analysis techniques, and therapeutic methods for addressing psychological conflicts demonstrates genuine empirical grounding in clinical practice (Jung, 1921; Shamdasani, 2003). The concept of psychological complexes—autonomous clusters of emotionally charged ideas—rests on observable clinical phenomena and has found validation in contemporary cognitive psychology through research on schemas and implicit memory (Jung, 1921; Greenwald & Banaji, 1995; Bargh & Chartrand, 1999).

Jung's method of "active imagination"—structured engagement with unconscious contents through imagery and fantasy—represents a significant therapeutic innovation that has influenced modern approaches to depth therapy (Jung, 1960). Clinical evidence suggests this technique can facilitate psychological insight and emotional processing, though the mechanism likely operates through established psychological processes rather than access to metaphysical unconscious structures.

7.1.2 The Collective Unconscious: Empirical Problems

Jung's concept of a collective unconscious containing universal archetypes faces severe empirical challenges that cannot be resolved through clinical observation alone (Jung, 1921; Shamdasani, 2003). While cross-cultural similarities in mythological motifs and psychological patterns exist, more parsimonious explanations are available through evolutionary psychology, cognitive constraints, and cultural diffusion processes (Boyer, 2001; Barrett, 2004).

Alternative explanations for archetypal patterns include cognitive architecture shaped by evolutionary pressures, environmental constraints generating convergent cultural solutions, developmental universals creating similar symbolic expressions, and cultural transmission spreading mythological motifs across cultures. Contemporary cognitive science provides robust frameworks for understanding psychological universals without requiring a metaphysical collective unconscious. Research on moral psychology, facial recognition, and language acquisition demonstrates how evolved cognitive modules can produce cross-cultural similarities through naturalistic mechanisms (Haidt, 2001).

7.1.3 Synchronicity: The Problem of Meaningful Coincidence

Jung's synchronicity concept—"meaningful coincidences" lacking causal connection—represents his most problematic theoretical innovation (Jung, 1952; Main, 2007). The famous scarab beetle case, while clinically interesting as a moment of therapeutic breakthrough, cannot support broader claims about acausal connections between psychological and physical events (Jung, 1952).

Cognitive explanations for synchronistic experience include pattern detection bias, where humans are evolutionarily predisposed to detect meaningful patterns, often perceiving significance in random events (Whitson & Galinsky, 2008). Additional factors include confirmation bias, where memorable coincidences are noted while non-coincidences are ignored (Nickerson, 1998), availability heuristic effects making striking events seem more probable than they actually are (Tversky & Kahneman, 1973), and retrospective meaning-making where significance is attributed to events after they occur rather than predicted beforehand.

The collaboration with Wolfgang Pauli, while historically fascinating, did not produce scientifically viable theories. Pauli's own later reservations about synchronicity reflect his recognition that the concept lacked the precision and testability required for scientific validity (Meier, 2001).

7.2 BOHM'S PHYSICS: LEGITIMATE SCIENCE AND SPECULATIVE EXTENSIONS

7.2.1 The de Broglie-Bohm Theory: Scientific Contribution

Bohm's most lasting contribution to physics lies in his development of the de Broglie-Bohm interpretation of quantum mechanics, which provides a mathematically rigorous alternative to standard interpretations while maintaining deterministic causation through hidden variables (Bohm, 1952a, 1952b; Holland, 1993). This work represents legitimate physics that has gained respect within the quantum foundations community, though it remains a minority interpretation.

The theory's significance lies not in any empirical advantages over standard quantum mechanics—it makes identical predictions—but in demonstrating that deterministic interpretations of quantum phenomena remain mathematically viable (Goldstein, 2013; Dürr, Goldstein, & Zanghì, 1992). This contribution to the conceptual foundations of physics stands independently of Bohm's later philosophical speculations.

7.2.2 The Implicate Order: From Physics to Metaphysics

Bohm's concept of the "implicate order"—an underlying reality where all parts are enfolded within each other—represents a transition from rigorous physics to speculative metaphysics. While philosophically interesting, the implicate order makes no testable predictions that distinguish it from conventional quantum mechanics and therefore cannot be evaluated through normal scientific criteria.

Bohm himself acknowledged this limitation, describing the implicate order as a "new mode of description" rather than a physical theory proper. The concept functions more as a metaphysical framework for interpreting existing physics than as a contribution to physics itself. This distinction is crucial: while scientists are free to hold metaphysical views, such views cannot claim scientific validation without empirical support.

7.2.3 Dialogue Method: Unvalidated Claims

Bohm's dialogue methodology, developed through collaboration with Jiddu Krishnamurti, represents an attempt to apply insights about wholeness and participation to group communication processes (Bohm, 1996). While this work has influenced organizational development and group facilitation practices, claims about its unique effectiveness lack systematic validation.

Methodological problems include absence of control groups, where most reports of dialogue effectiveness lack comparison with alternative group processes. Success is often measured through subjective evaluation criteria using participant self-reports rather than objective outcomes. Benefits may result from facilitator effects rather than the specific dialogue method. Selection bias may also play a role, as groups choosing to engage in dialogue may be predisposed toward collaborative outcomes.

Contemporary research on group decision-making and collaborative problem-solving has not demonstrated superior outcomes for Bohmian dialogue compared to other structured group processes when evaluated through controlled studies.

7.3 THE PROBLEM OF CONCEPTUAL CONVERGENCE

7.3.1 Linguistic Similarity versus Theoretical Alignment

The apparent convergences between Jung and Bohm often rest on shared vocabulary rather than equivalent concepts. Terms like "wholeness," "participation," and "process" function differently within their respective theoretical frameworks and cannot be meaningfully unified without conceptual confusion.

The term "wholeness" has different referents: for Jung, it refers to psychological coordination of conscious and unconscious contents through individuation, while for Bohm, it describes physical undivided reality described by the implicate order. Similarly, "participation" involves different mechanisms: Jung's psychological engagement with unconscious archetypal contents versus Bohm's observer effects in quantum measurement and dialogue processes. The concept of "process" operates on different temporalities: Jung's lifelong psychological development through conflict resolution versus Bohm's moment-to-moment unfolding of implicate order into explicate phenomena.

These differences reflect deeper incompatibilities between psychological and physical domains that cannot be resolved through terminological convergence alone.

7.3.2 Category Errors in Coordination Attempts

Attempts to synthesize Jungian psychology with Bohmian physics commit systematic category errors by applying concepts appropriate to one domain within a different domain with different validation criteria and explanatory targets.

Psychological concepts are misapplied to physics when quantum measurement is treated as analogous to psychological participation in archetypal contents, quantum nonlocality is interpreted through Jungian synchronicity concepts, or therapeutic insights are used to explain physical phenomena. Conversely, physical concepts are misapplied to psychology when psychological wholeness is treated as analogous to quantum entanglement, symbolic processes are interpreted through implicate order concepts, or quantum mechanics is used to validate claims about the collective unconscious.

These category errors obscure rather than clarify both psychological and physical phenomena by imposing inappropriate explanatory frameworks across domain boundaries.

7.4 CONTEMPORARY APPLICATIONS: CIRCUMSCRIBED CONTRIBUTIONS

7.4.1 Consciousness Studies: Limited Influence

Claims about Jung and Bohm's influence on contemporary consciousness studies require careful qualification. While both figures addressed questions about the nature of consciousness that remain relevant, their specific theoretical contributions have had limited impact on current empirical research.

Legitimate influences include Jung's emphasis on subjective experience as irreducible to neurobiological processes, which anticipates contemporary debates about the "hard problem" of consciousness. Bohm's attention to the role of observation in physical systems relates to ongoing discussions about quantum mechanics and consciousness. Both figures' holistic approaches have influenced integrative perspectives in consciousness studies.

However, overstated claims must be corrected. Integrated Information Theory developed independently of both figures through mathematical approaches to information processing (Tononi, 2008). Mainstream neuroscience research on meditation and contemplative practices draws primarily on neuroscientific and psychological frameworks. Quantum theories of consciousness remain highly speculative and controversial within neuroscience (Tegmark, 2000).

7.4.2 Practical Applications: Mixed Evidence

Applications of Jungian and Bohmian insights in therapeutic and organizational contexts show mixed results that must be evaluated carefully to separate effective techniques from theoretical commitments.

In therapeutic applications, Jungian therapeutic techniques, when evaluated through controlled studies, show modest effectiveness comparable to other depth therapies. Active imagination and dream analysis can facilitate psychological insight through established cognitive and emotional mechanisms (Jung, 1960). However, no evidence supports the necessity of accepting archetypal or collective unconscious theories for therapeutic effectiveness.

In organizational applications, Bohmian dialogue techniques may improve group communication through structured attention to assumptions and listening processes. Benefits likely result from general principles of effective facilitation rather than specific theoretical commitments. Systematic evaluation comparing dialogue methods with other group processes remains limited.

7.5 METHODOLOGICAL LESSONS: COORDINATION WITH SAFEGUARDS

7.5.1 Translation Protocols for Interdisciplinary Research

The Jung-Bohm case study illuminates the importance of developing rigorous translation protocols for interdisciplinary research that avoid category errors while enabling productive dialogue across domain boundaries.

Essential safeguards include domain specification to clearly identify which claims belong to psychology, physics, or philosophy. Distinct validation criteria must be maintained appropriate to each domain. Conceptual precision requires defining key terms operationally within each domain before attempting comparisons. Limitation acknowledgment explicitly identifies where analogies break down and coordination becomes impossible.

7.5.2 Heuristic Value versus Explanatory Claims

Both Jung and Bohm's contributions retain value as sources of heuristic insights for interdisciplinary research when properly circumscribed. Their emphasis on holistic thinking, attention to subjective experience, and participatory methodologies can inform research practices without requiring acceptance of their broader theoretical claims.

Legitimate heuristic applications include structured attention to assumptions and presuppositions in research processes, coordination of subjective reports with objective measurements in consciousness studies, development of collaborative research methodologies that honor multiple perspectives, and recognition of researcher participation in shaping research outcomes.

7.6 CONCLUSION: BOUNDED COORDINATION AND METHODOLOGICAL RIGOR

Carl Gustav Jung and David Bohm made significant contributions within their respective domains that continue to offer valuable insights for contemporary research. Jung's clinical innovations and attention to subjective experience provide important resources for therapeutic practice and consciousness studies, while Bohm's quantum foundations work and dialogue methodology offer legitimate contributions to physics and collaborative inquiry.

However, attempts to synthesize their work into unified theories of "participatory consciousness" face insurmountable methodological problems stemming from category errors, different validation criteria, and speculative extensions beyond empirical support. Their apparent convergences often reflect linguistic similarity rather than theoretical alignment, and their most ambitious integrative claims cannot meet the evidence standards required for scientific validity.

The most defensible approach to their legacy involves preserving domain-specific contributions by maintaining Jung's clinical insights within psychological contexts and Bohm's physics contributions within appropriate scientific frameworks. This requires rejecting premature synthesis and avoiding attempts to force coordination across incompatible domains with different validation criteria. Methodological insights can be extracted to identify heuristic principles that inform interdisciplinary research without requiring acceptance of broader theoretical claims. Translation protocols must be developed to create rigorous safeguards for interdisciplinary dialogue that prevent category errors while enabling productive exchange.

This approach maintains intellectual honesty about the limitations of coordination while preserving the legitimate contributions both figures made to our understanding of consciousness, wholeness, and participatory inquiry. Their work demonstrates both the promise and the pitfalls of ambitious interdisciplinary synthesis, providing valuable lessons for contemporary researchers committed to bridging disciplinary boundaries without sacrificing methodological rigor.

The ultimate significance of the Jung-Bohm case lies not in validating their specific theoretical claims, but in illuminating the methodological requirements for responsible interdisciplinary research that honors both the integrity of distinct domains and the creative possibilities of carefully bounded coordination.

CHAPTER 8: PATTERNS IN KNOWLEDGE COORDINATION - HEURISTIC PRINCIPLES AND METHODOLOGICAL CONSTRAINTS

ABSTRACT

The examination of twelve historical figures across diverse intellectual traditions reveals recurring methodological patterns that warrant systematic analysis. However, the identification of such patterns requires careful attention to the limitations of our evidence base and the constraints inherent in comparative intellectual history. This chapter extracts four heuristic principles from the historical cases while maintaining explicit boundaries around their scope and applicability.

The sample examined—Pythagoras, Plato, Avicenna, Al-Biruni, Aquinas, Leonardo, Leibniz, Goethe, Teilhard, Iqbal, Jung, and Bohm—represents a particular trajectory of intellectual history heavily weighted toward Western and Islamic philosophical traditions. This selection reflects both the availability of translated primary sources and the scholarly focus on figures already recognized for synthetic approaches. Such constraints necessarily limit the universalizability of any patterns identified and require explicit acknowledgment of cultural and methodological biases.

The analytical framework employed here distinguishes between pattern recognition as a heuristic exercise and empirical validation of causal relationships. While computational text analysis can identify semantic clusters and conceptual networks, the interpretation of these patterns remains fundamentally hermeneutic rather than statistical (Blei et al., 2003). Claims about effectiveness or success rates require controlled comparative studies that lie beyond the scope of historical analysis.

8.1 METHODOLOGICAL CONSIDERATIONS AND VALIDATION CONSTRAINTS

8.1.1 Pattern Recognition and Its Limitations

The identification of cross-temporal patterns in intellectual history faces several methodological challenges that constrain interpretive confidence. Selection bias represents perhaps the most significant limitation: by examining figures already recognized for integrative work, we inevitably find integrative patterns (Ioannidis, 2005). The many scholars who achieved significant advances through specialization, reduction, or domain-specific focus remain absent from this analysis, potentially skewing our understanding of what enables intellectual progress.

Cultural translation poses equally serious challenges. When we identify apparent similarities between concepts like Pythagorean harmonia, Avicenna's ittihad, and Bohm's "implicate order," we risk imposing false equivalences across distinct metaphysical frameworks. These terms emerge from different linguistic, philosophical, and cultural contexts that may share surface features while differing fundamentally in their systematic roles and implications.

Retrospective interpretation introduces additional distortions. Contemporary categories of "empirical," "rational," "intuitive," and "participatory" knowledge may not map meaningfully onto historical epistemic practices. The danger of anachronism—reading present concerns into past texts—requires constant vigilance and explicit acknowledgment of interpretive limitations (Skinner, 1969).

8.1.2 Validation Criteria and Evidence Standards

The patterns identified here rely primarily on textual analysis and scholarly interpretation rather than rigorous empirical validation. While this approach can illuminate conceptual relationships and methodological similarities, it cannot establish causal claims about effectiveness or predict success in contemporary applications.

Three validation constraints deserve particular emphasis:

Historical Evidence: Claims about what historical figures "achieved" or how their methods "worked" depend on contested interpretations of their contributions and influence.

Causal Attribution: Apparent correlations between integrative methods and intellectual success may reflect survivorship bias rather than causal relationships (Ioannidis, 2005).

Contemporary Relevance: Applications of historical patterns to current problems require independent validation through controlled studies rather than analogical reasoning alone.

8.2 FOUR HEURISTIC PRINCIPLES: PATTERNS AND LIMITATIONS

8.2.1 Multi-Dimensional Epistemology (MDE)

Pattern Identification: The historical figures examined typically coordinated multiple sources of knowledge rather than relying exclusively on single epistemic modes. This includes various combinations of empirical observation, logical analysis, intuitive insight, textual interpretation, and experiential engagement.

Historical Examples: Plato's Republic presents different levels of knowing from conjecture to dialectical understanding; Avicenna integrates Aristotelian demonstration with mystical insight; Leonardo combines direct observation with mathematical analysis and artistic representation; Goethe develops "delicate empiricism" coordinating sensory observation with imaginative participation.

Limitations and Counterexamples: Many significant intellectual advances have emerged through single-mode focus. The logical positivists' emphasis on empirical verification, while ultimately limited, produced valuable methodological clarifications. Pure mathematics has generated crucial insights through abstract reasoning divorced from empirical application. The apparent prevalence of multi-dimensional approaches in our sample may reflect selection bias rather than necessity (Ioannidis, 2005).

Contemporary Applications: Some current research programs appear to benefit from methodological coordination—climate science combining computational modeling with local ecological knowledge, consciousness studies integrating neuroscientific measurement with phenomenological reports, design thinking coordinating analytical and creative processes (Thompson, 2007; Varela et al., 1991). However, systematic comparison with single-mode approaches would be required to establish superior effectiveness.

8.2.2 Hierarchical-Organic Organization (HOO)

Pattern Identification: Many figures in our sample organize knowledge through hierarchical levels that maintain organic rather than merely aggregative relationships. Higher levels emerge from but cannot be reduced to lower levels, while exhibiting both upward and downward causal influences.

Historical Examples: Aquinas's Summa Theologiae articulates participation metaphysics where each level participates analogically in higher levels while maintaining proper perfection; Leibniz's Monadology presents infinite hierarchical nesting of reality; Goethe's morphological studies trace transformations of fundamental forms across organizational levels.

Alternative Approaches: Many successful knowledge frameworks explicitly reject hierarchical organization. Network theories, rhizomatic philosophies, and various Indigenous knowledge systems achieve coordination through lateral connections rather than vertical stratification (Barabási, 2016; Ford et al., 2010). The prevalence of hierarchy in our sample may reflect historical and cultural biases rather than functional necessity.

Contemporary Relevance: Complexity science provides mathematical frameworks for understanding hierarchical emergence in natural systems (Mitchell, 2009; Levin, 1998). However, many successful contemporary integrative approaches—network analysis, systems thinking, ecological modeling—operate through non-hierarchical principles with comparable effectiveness.

8.2.3 Process-Relational Ontology (PRO)

Pattern Identification: Several figures emphasize process, temporality, and relational constitution over substance metaphysics, treating entities as relatively stable patterns within flux rather than independent objects with intrinsic properties (Whitehead, 1978).

Historical Examples: Al-Biruni describes reality through samsara (continuous flow) integrated with Aristotelian dynamics; Teilhard's evolutionary mysticism makes becoming primary to being; Iqbal presents reality as "pure duration"; Bohm's implicate order emphasizes temporal unfolding over static structure.

Static Alternatives: Prominent integrators like Parmenides and Spinoza achieved sophisticated syntheses while emphasizing permanence over process. Mathematical Platonism integrates diverse domains through timeless abstract objects (Balaguer, 1998). Many successful scientific theories operate through equilibrium states and conservation laws rather than processual dynamics.

Domain Specificity: Process thinking appears particularly valuable for understanding developmental, evolutionary, and historical phenomena. However, many areas of successful knowledge coordination—logical systems, mathematical structures, physical constants—benefit from static rather than processual approaches.

8.2.4 Participatory Consciousness (PC)

Pattern Identification: Many figures treat knowledge as emerging through participatory engagement between knower and known rather than detached observation of independent objects. This involves the co-constitution of subject and object through the knowing act.

Historical Examples: Plotinus describes intellectual knowledge through henosis (union) where knower becomes identified with the known; Sufi ma'rifa involves participatory knowledge where "the knower, the known, and the knowledge become one"; Goethe's "exact sensorial imagination" requires developing new perceptual capacities; Jung's active imagination engages psychic contents as autonomous partners.

Formal Alternatives: Many highly successful integrative approaches achieve synthesis through systematic abstraction and formal modeling that explicitly brackets participatory elements. Mathematical coordination across domains, computational modeling, and algorithmic analysis often work best when minimizing subjective participation rather than embracing it.

Validation Challenges: Claims about participatory knowledge face particular difficulties in validation. While first-person reports of transformative knowing experiences are phenomenologically interesting, establishing their epistemic value compared to non-participatory approaches requires controlled comparative studies.

8.3 INTERRELATIONSHIPS AND SYSTEMATIC CONSIDERATIONS

The four patterns identified exhibit potential complementary relationships, though their interaction remains empirically unestablished. Multi-dimensional epistemology might enable recognition of different organizational levels (hierarchical-organic), which could manifest as temporal processes (process-relational) accessible through engaged participation (participatory consciousness).

However, this apparent systematic coherence may reflect interpretive bias rather than functional necessity. Each principle faces counterexamples and alternative approaches that achieve comparable results through different means. The appearance of systematic coordination may result from post-hoc rationalization rather than genuine methodological requirements.

Contemporary applications suggest that selective rather than comprehensive deployment of these principles often proves most effective. Climate science benefits from multi-dimensional knowledge coordination and process thinking but may not require hierarchical organization or participatory consciousness. Neuroscience gains from methodological coordination and hierarchical modeling while maintaining observational distance rather than participatory engagement.

8.4 CONTEMPORARY APPLICATIONS: ILLUSTRATIVE CASES

8.4.1 Climate Science Coordination

Climate research demonstrates apparent benefits from multi-dimensional epistemology through the coordination of computational models, observational data, paleoclimatic evidence, and indigenous ecological knowledge (Edwards, 2010; Bojinski et al., 2014). Process-relational thinking proves essential for understanding Earth system dynamics and feedback loops. However, the field's success may result from empirical necessity rather than methodological principle—complex environmental systems simply require multiple data sources and processual thinking for adequate understanding.

The incorporation of indigenous knowledge systems represents genuine methodological innovation, though systematic evaluation of its effectiveness compared to purely computational approaches remains incomplete (Ford et al., 2010). While participatory approaches increasingly involve stakeholder communities in research design, this may reflect political and ethical requirements rather than epistemic advantages.

8.4.2 Consciousness Studies

Contemporary consciousness research exhibits methodological coordination across neuroscientific, psychological, and phenomenological approaches. The field increasingly recognizes that first-person experiential reports provide irreducible data for understanding conscious states, suggesting potential validation of participatory approaches.

However, the most rigorous advances in consciousness studies—

mathematical theories like Integrated Information Theory, computational models of attention and working memory, neuroscientific discoveries about brain networks—typically emerge through conventional scientific methods rather than participatory epistemologies (Tononi, 2008; Dehaene, 2014). The role of first-person approaches may remain supplementary rather than foundational.

8.4.3 Design and Innovation

Design thinking explicitly coordinates analytical problem-solving with creative ideation, stakeholder engagement, and iterative prototyping (Cross, 2011; Dorst, 2011). This methodological coordination appears to generate innovations unavailable through purely analytical or purely creative approaches.

Yet the systematic evaluation of design thinking's effectiveness remains limited. Many successful innovations emerge through serendipity, systematic research and development, or market-driven iteration rather than integrative design methodologies. The apparent success of design thinking may reflect selection bias in reported cases rather than superior effectiveness (Ioannidis, 2005).

8.5 CRITICAL ASSESSMENT AND BOUNDARY CONDITIONS

8.5.1 Addressing Objections

Specialization and Reduction: The most serious challenge to integrative approaches comes from the documented success of specialization and reduction across multiple domains. The discovery of DNA's structure, the development of quantum mechanics, and advances in molecular biology typically resulted from focused expertise rather than broad coordination (Watson & Crick, 1953; Kragh, 1999). Many contemporary scientific breakthroughs continue emerging through deep specialization within narrow domains.

Response: Coordination appears most valuable at the interfaces between established specialties rather than as a replacement for specialized depth. The principles identified here may function best as coordination mechanisms after specialized knowledge has been developed rather than as alternatives to rigorous domain-specific work.

Cultural Imperialism: The heavy Western and Islamic representation in our sample raises legitimate concerns about imposing particular cultural frameworks on diverse knowledge traditions. The very categories of "empirical," "rational," "intuitive," and "participatory" may reflect specifically Western epistemic distinctions.

Response: This analysis acknowledges its cultural limitations and presents patterns as heuristics rather than universal principles. Alternative frameworks from other traditions—Chinese correlative thinking, Indigenous holistic approaches, African communal epistemologies—may reveal entirely different integrative patterns equally valid within their contexts.

Confirmation Bias: By selecting figures known for coordination, we inevitably find integrative patterns, potentially creating circular reasoning where we "discover" what we already assumed.

Response: This limitation is acknowledged explicitly, and the patterns are presented as observations about a particular intellectual lineage rather than universal features of knowledge production. Future research should examine figures explicitly committed to specialization or reduction to provide comparative perspective.

8.5.2 Boundary Conditions and Appropriate Applications

The heuristic principles identified here appear most applicable under specific conditions rather than universally:

Problem Complexity: Coordination becomes valuable when problems span multiple domains or scales requiring different types of expertise and evidence.

Stakeholder Diversity: Participatory approaches gain importance when multiple communities hold relevant knowledge or will be affected by research outcomes.

Temporal Dynamics: Process thinking becomes essential when change, development, or feedback loops represent crucial phenomena.

Scale Coordination: Hierarchical organization proves useful when phenomena operate across multiple organizational levels with emergent properties.

When these conditions are absent, specialized approaches may prove more effective than integrative ones. The principles should be deployed selectively based on problem characteristics rather than applied universally.

8.6 CONCLUSION: HEURISTIC VALUE AND METHODOLOGICAL HUMILITY

The four patterns identified—multi-dimensional epistemology, hierarchical-organic organization, process-relational ontology, and participatory consciousness—represent recurring themes in a specific sample of historical integrators rather than universal principles of knowledge coordination. Their value lies in their heuristic potential for addressing contemporary problems that span disciplinary boundaries, involve multiple stakeholders, or require coordination across different types of evidence.

However, this potential remains largely untested through systematic comparative studies. Claims about their effectiveness require empirical validation rather than historical argumentation. Many successful knowledge enterprises operate through specialization, reduction, formal abstraction, and detached observation—approaches that explicitly reject the principles identified here.

The path forward requires methodological pluralism rather than integrative orthodoxy (Longino, 2002; Kellert et al., 2006). Some problems may benefit from integrative approaches guided by these historical patterns, while others may require focused specialization or alternative methodologies not represented in this sample. The ultimate criterion should be pragmatic effectiveness rather than theoretical elegance or historical precedent.

Future research should expand beyond the Western-Islamic trajectory examined here to include other cultural traditions of knowledge coordination. Equally important is a systematic study of successful specialization and reduction to understand when coordination adds value versus when it introduces unnecessary complexity or bias.

The historical figures examined here demonstrate that knowledge coordination remains possible across disciplinary boundaries and cultural contexts. Their approaches provide tested heuristics for contemporary challenges requiring synthesis across diverse domains. However, their legacy should inspire methodological experimentation rather than dogmatic application, creative adaptation rather than mechanical replication.

The fragmentation of knowledge represents a genuine contemporary challenge requiring innovative responses. The historical patterns identified here offer one set of tools for addressing this challenge, but they remain tools rather than solutions, heuristics rather than algorithms, invitations to experiment rather than prescriptions to follow.

GLOSSARY

Al-Biruni (Abū Rayḥān al-Bīrūnī)
A medieval Islamic scholar who advanced knowledge coordination through an empirical-comparative methodology. His approach involved a deep linguistic immersion in foreign traditions, such as learning Sanskrit, to achieve a nuanced "understanding from within" before commencing systematic comparison. He used mathematics as a neutral, precise language to bridge cultural differences, as demonstrated by his work in synchronizing various calendrical systems. This method offers a powerful model for cross-cultural collaboration, emphasizing how coordination can be achieved without forcing a premature or literal unification of disparate knowledge systems. The book highlights his use of comparative matrices, which allowed him to preserve and articulate deep differences between traditions while still enabling productive dialogue.

Analogical Predication
A coordination tool developed by Thomas Aquinas to address the fundamental problem of how concepts and language could be applied across radically different levels of reality, such as the finite and the infinite. The theory of analogy provided a principled way to relate these disparate domains without either reducing them to one another or treating them as completely unrelated. This enabled the systematic coordination of abstract conceptual analysis with concrete ontological commitment.1

Analytical Psychology
A school of depth psychology founded by Carl Gustav Jung. It is examined in the book as a historical experiment in interdisciplinary coordination, particularly for its attempts to bridge psychological phenomena with concepts from theoretical physics. The analysis in the book draws a distinction between Jung's clinically grounded and empirically defensible therapeutic methods, such as active imagination, and his more speculative theoretical claims, which lack robust empirical support.

Aquinas (Thomas)
A medieval philosopher and theologian who is presented as a case study in systematic knowledge coordination within theological boundaries. He developed a set of explicit procedures, such as the Disputed Question Method and Analogical Predication, to handle apparent conflicts between philosophical reasoning and theological commitment. The book notes that his apparent synthesis was enabled by, and inextricably linked to, the institutional context of the medieval university and

its ecclesial constraints, which ultimately subordinated philosophical inquiry to theological authority.

Avicenna (Ibn Sīnā)

A medieval Islamic scholar who attempted knowledge coordination through the creation of a systematic philosophical architecture. He employed the essence-existence distinction and an emanationist framework to organize disparate knowledge domains. The book highlights his work as an example of how coordination can be a creative act of translation, noting that the philosophical vocabulary he developed (e.g., *wujūd* and *māhiyya*) enabled new conceptual distinctions that influenced both Islamic and European thought. This terminological innovation facilitated coordination across frameworks but also entailed a conceptual transformation of the original source material, demonstrating that translation is not a neutral process.

Biomimetic Investigation

A coordination strategy used by Leonardo da Vinci that involved the systematic study of natural forms as a source for technical innovation and engineering design. Leonardo's studies of bird flight, water dynamics, and plant structures are presented as examples of how he productively coordinated insights from biological observation with the practical requirements of engineering, creating a genuine cross-domain learning process that avoided both reductive mechanism and vague organicism.

Boundary Objects

Shared resources or artifacts that enable interdisciplinary collaboration by functioning as a common point of reference for different knowledge communities without forcing those communities to abandon their distinct methodological commitments.1 Examples provided in the book include standardized datasets, measurement protocols, and conceptual bridges, such as the data provided by the Global Climate Observing System, that can be used by atmospheric physicists and policy analysts alike. The development of boundary objects is framed as an operational solution to the problem of fragmentation, allowing for coordination without requiring homogenization of diverse fields.

Characteristica Universalis

A project conceived by Gottfried Wilhelm Leibniz to create a universal symbolic language capable of expressing all possible thoughts with such precision that intellectual disputes could be resolved through calculation rather than argumentation. This project is a prime example of formal coordination and represents an ambitious, though ultimately unrealized, attempt at comprehensive theoretical unification. The book notes that while the project itself failed, Leibniz's work on symbolic logic and notation proved to be a durable and influential contribution.

Collective Unconscious

A concept in Jungian psychology that proposes the existence of an unconscious containing universal archetypes and psychological patterns shared by all humanity. The book argues that this concept, while appearing to provide a coordinating framework for human experience, faces severe empirical challenges.1 The document explains that more parsimonious, naturalistic explanations for cross-cultural similarities—such as evolutionary cognitive architecture, shared environmental constraints, and cultural transmission—can account for the same phenomena without recourse to a metaphysical unconscious.

Complexity-Consciousness Hypothesis

A central claim by Teilhard de Chardin proposing a fundamental correlation between increasing material complexity and enhanced conscious experience throughout evolutionary history. The book's critical assessment of this hypothesis points out its fundamental problems, including a lack of sufficiently precise definitions for "complexity" and "consciousness" to enable empirical testing, the existence of counterexamples in nature, and a noticeable anthropocentric bias that appears to privilege traits leading to human cognition.

Consilience

A term popularized by Edward O. Wilson, which refers to the project of reducing all valid knowledge to fundamental physical laws through hierarchical explanatory chains. The book presents this as a failed but instructive coordination strategy, arguing that it encounters fundamental obstacles because emergent properties at higher organizational levels resist a complete explanation through lower-level mechanisms.1 The project's limitations highlight the difficulty of achieving true unification and underscore the value of more pragmatic coordination strategies.1

Coordination Without Unification

A crucial distinction that underpins the book's entire analytical framework. This approach aims to improve interfaces between heterogeneous knowledge domains without eliminating their fundamental differences. It is presented as a more pragmatic and immediately valuable goal than the comprehensive theoretical synthesis that unification seeks. The book suggests that civilizational challenges like climate change can be addressed by coordinating insights from different fields, such as atmospheric physics, economics, and political science, without first requiring theoretical unification across those domains, acknowledging that some fields may be irreducibly different.

De Broglie-Bohm Theory

A mathematically rigorous interpretation of quantum mechanics developed by David Bohm. The book identifies this as his "most lasting contribution to physics" and highlights its significance for demonstrating that deterministic interpretations of quantum phenomena remain mathematically viable, even if they lack empirical advantages over other interpretations.1 This work is explicitly distinguished from Bohm's later philosophical speculations, serving as a model for separating legitimate scientific contributions from speculative extensions.

Dialectical Architecture

A term used in the book to describe Plato's systematic philosophical methodology for organizing knowledge. This architecture includes the Divided Line as an organizational framework for different modes of inquiry and the method of division and collection for identifying both commonalities and differences across domains. This approach provides a transferable principle for contemporary coordination: the systematic investigation of where unity exists and where differences persist, rather than simply asserting them.

Disputed Question Method (Quaestio Disputata)

A procedure used by Thomas Aquinas for systematic coordination within theological boundaries. This format enabled the handling of apparent conflicts between different authorities and methods by explicitly presenting objections, citing contrary authorities, and providing a reasoned response. This procedural coordination provided a transparent way of relating different types of evidence and argument, though it operated within predetermined doctrinal boundaries that ultimately limited its openness.

Divided Line

A passage in Plato's Republic that provides a methodological scaffold for organizing different types of cognitive engagement with reality. The book interprets the Divided Line not as metaphysical doctrine but as a framework that clarifies how different warrant domains—from empirical observation to mathematical reasoning and philosophical dialectic—relate hierarchically while maintaining their autonomous validity.

Emanationist Framework

A cosmological system, used by Avicenna, that coordinates divine simplicity with cosmic complexity through hierarchical principles. In this framework, the Necessary Existent emanates existence through successive intelligences, structuring relationships across metaphysical, cosmological, and psychological domains. The book notes that this architectural sensibility is also evident in Avicenna's medical work, where he organized theoretical foundations with clinical practice to enable the durable transfer of knowledge across cultures.

Empirical-Aesthetic Coordination

A coordination strategy pioneered by Leonardo da Vinci that integrated observational investigation, technical problem-solving, and aesthetic representation. Leonardo's method of "Sapere Vedere" is the central principle of this approach, allowing him to coordinate empirical investigation with artistic representation in his detailed anatomical and biomimetic studies. This approach highlights the value of visual and embodied coordination as a means of bridging different knowledge domains.

Essence–Existence Distinction

A coordination tool developed by Avicenna that enabled the systematic analysis of how particulars relate to general categories without reducing them to either Platonic forms or Aristotelian substantial forms. This conceptual innovation is a key example of how terminological precision can facilitate coordination by providing a framework for analyzing logical and metaphysical relationships.

Evolutionary Coordination

A term used to describe the ambitious attempts by figures like Pierre Teilhard de Chardin and Muhammad Iqbal to reconcile evolutionary science with religious thought. The book's critical assessment concludes that while their efforts were historically significant, their specific syntheses were fundamentally problematic, largely due to their commitment to a teleological view of evolution that conflicts with modern scientific understanding.

Evolutionary Theism

The philosophical framework developed by Muhammad Iqbal, which proposed that divine creative action operates through evolutionary processes rather than in spite of them. Iqbal's concept of a "dynamic selfhood" suggested that consciousness at all levels participates in cosmic creativity, with evolution representing the temporal manifestation of divine purpose.[1] The book argues that this framework runs into problems of theological inconsistency and causal overdetermination.

Formal Coordination

A coordination strategy pursued by Gottfried Wilhelm Leibniz that aimed for comprehensive unification through the development of a universal symbolic language and a systematic metaphysics. This approach is contrasted with the phenomenological coordination of Goethe, and the book notes that while Leibniz's ambitions were grand, they ultimately failed due to theoretical overreach and a lack of institutional viability.

Forms Theory
A concept from Plato that posits transcendent, eternal structures that particular phenomena instantiate. The book suggests that while the theory has well-known metaphysical problems, it contains methodologically valuable insights about structural thinking. The emphasis on formal organization over material substance anticipates contemporary approaches in information theory and complexity science.

Fragmentation
Defined operationally in the book as observable interface problems rather than a civilizational pathology. The document provides specific, measurable examples of fragmentation, including the divergence of technical vocabularies (translation costs), the incommensurability of data standards (format incompatibilities), and the dominance of certain fields in policy discussions (power asymmetries). By defining the problem in this way, the book reframes it as a solvable design challenge rather than an insurmountable crisis.

Goethe (Johann Wolfgang von)
A figure who pursued phenomenological coordination through his morphological method and phenomenological empiricism. The book presents his work as an ambitious, but ultimately failed, project of systematic synthesis, noting that his approach, which prioritized qualitative observation over quantitative measurement, faced significant institutional resistance from the emerging scientific community.

Hierarchical-Organic Organization (HOO)
One of the four heuristic principles identified in the book which involves creating a hierarchical structure that preserves disciplinary integrity while enabling interface. The "organic" aspect of this principle means that higher levels emerge from but cannot be reduced to lower ones, exhibiting both upward and downward causal influences. Historical examples, such as Aquinas's *Summa Theologiae*, demonstrate how this method can clarify relationships between different domains, but the book cautions that such hierarchies in the past often embedded value judgments that are not appropriate for contemporary pluralistic contexts.

Implicate Order

A speculative concept from David Bohm describing an underlying reality where all parts are enfolded within each other. The book argues that this concept functions more as a metaphysical framework for interpreting existing physics than as a testable scientific theory. The document's analysis shows that the concept makes no testable predictions that distinguish it from conventional quantum mechanics, making it unsuitable for scientific evaluation and highlighting the crucial distinction between a philosophically interesting idea and a scientifically valid one.

Iqbal (Muhammad)

A modern figure who attempted evolutionary coordination by proposing an evolutionary theism that integrated reason, empirical experience, and intuitive insight from an Islamic perspective. The book's critical assessment of his work, alongside that of Teilhard, points to the "Orthogenesis Problem" as a core weakness in his attempt to reconcile a teleological view of evolution with modern science.

Jung (Carl Gustav)

A prominent figure who developed analytical psychology and concepts like the Collective Unconscious and Synchronicity. The book examines his work for its limits and possibilities in interdisciplinary coordination, drawing a clear distinction between his empirically grounded clinical methods and his more problematic theoretical innovations, which often rely on concepts that can be more parsimoniously explained by cognitive biases or other naturalistic mechanisms.

Knowledge Coordination

The central concept of the book, defined as the creation of durable coordination mechanisms between heterogeneous knowledge communities. The book frames this as a problem of interface design, a pragmatic approach that contrasts with the quest for metaphysical unity or a nostalgic return to a golden age of unified wisdom. The book analyzes historical figures as experiments in this design process, seeking to identify transferable principles and boundary conditions for effective collaboration.

Leonardo da Vinci

A Renaissance polymath who is presented as a case study in empirical-aesthetic coordination through visual investigation. His approach, based on the principle of "Sapere Vedere" and biomimetic investigation, enabled him to integrate direct observation with mathematical and aesthetic principles.1 The book notes that while his achievements were remarkable, his coordination method remained largely personal rather than institutional and lacked the systematic protocols that would allow for broader replication and dissemination.

Leibniz (Gottfried Wilhelm)
A prominent figure who pursued ambitious coordination projects through formal, logical unification. His two main projects, the
Characteristica Universalis and his Monadological Metaphysics are presented as experiments in formal coordination that, while ultimately failing in their grand ambitions, still produced valuable methodological innovations, such as the development of symbolic logic and a systematic approach to analogical reasoning. The book uses his work to illustrate the difficulty of achieving comprehensive theoretical unification.

Monadological Metaphysics
A systematic metaphysics developed by Leibniz that sought to ground all phenomena in simple substances, or monads, coordinated through a pre-established harmony. The book notes that this system, while ambitious in its attempt to preserve both mechanistic regularity and qualitative experience, ultimately depended on theological assumptions and created new philosophical problems that limited its empirical viability.

Morphological Method
A method developed by Goethe for investigating plant and animal forms.[1] This approach sought to discover systematic transformation principles through sustained, disciplined observation, with the goal of identifying underlying patterns (*Urphänomene*) manifest in biological variation. The book notes that while this method yielded genuine insights, it was largely qualitative and lacked the quantitative rigor to compete with the emerging mathematical approaches of his time.

Multi-Dimensional Epistemology (MDE)
One of the four heuristic principles identified in the book. It involves the systematic integration of multiple sources of knowledge, such as empirical observation, logical analysis, and intuitive insight. This principle is particularly applicable to problems that require different types of expertise and evidence, such as in climate science, where computational modeling is coordinated with indigenous ecological knowledge.

Natural Law
A concept used by Thomas Aquinas for cross-domain coordination, which enabled him to integrate ethics, politics, jurisprudence, and theology within a single, hierarchical framework. The book notes that this method, while effective for its time,

embedded specific value commitments about divine authority and social hierarchy that would be difficult to transfer to contemporary pluralistic contexts without significant adaptation.

Orthogenesis Problem

The fallacy of assuming that evolution exhibits an inherent directionality or a teleological tendency toward greater complexity or perfection. The book identifies this as a fundamental problem in the evolutionary coordination attempts of Teilhard de Chardin and Muhammad Iqbal. The document explains that this assumption conflicts fundamentally with the modern understanding of natural selection, which is a local and contingent process, not a goal-oriented one. This failure serves as a cautionary tale about the limits of philosophical overreach in scientific domains.

Participatory Consciousness (PC)

One of the four heuristic principles identified in the book, which proposes that knowledge emerges through a participatory engagement between the knower and the known, rather than from detached observation. The book notes that while this approach, exemplified by figures like Goethe and Jung, has heuristic value for generating hypotheses, its claims often lack the empirical testability required for scientific validation. The analysis shows that this principle is valuable for informing research and acknowledging researcher situatedness, but it must be supplemented with rigorous validation criteria from other domains.

Phenomenological Coordination

A coordination strategy pursued by Johann Wolfgang von Goethe that relied on a morphological method and phenomenological empiricism. This approach sought systematic principles through a disciplined engagement with natural phenomena, emphasizing qualitative observation over formal theorization. The book contrasts this with Leibniz's formal coordination, noting that while Goethe's approach yielded valuable insights about organic unity, its lack of quantitative rigor limited its acceptance within the scientific community.

Phenomenological Empiricism

A methodological principle developed by Goethe that attempted to integrate rigorous observation with aesthetic perception through disciplined phenomenological investigation. This approach recognized the qualitative dimensions of experience that mechanical observation alone could not capture, and the book notes that it remains relevant for certain types of qualitative research today.

Process-Relational Ontology (PRO)

One of the four heuristic principles identified in the book, which emphasizes process, temporality, and dynamic relationships over static categories. This approach treats entities as relatively stable patterns within a dynamic flux rather than as fixed, independent objects.1 The book notes that this kind of thinking is particularly valuable for understanding developmental and evolutionary phenomena, but that it is not universally applicable to all domains of knowledge.

Productive Specialization
A key concept in the book that acknowledges that the fragmentation of knowledge is not a symptom of intellectual decline but a predictable consequence of the success of specialized expertise. The document argues that this specialization has led to remarkable achievements and that the goal of knowledge coordination is to manage the interface problems it creates, not to abandon specialization itself.

Pythagoras and Plato
Early Greek thinkers who are examined as historical experiments in knowledge coordination through mathematical structure. The book analyzes their attempts to use formal patterns—such as the simple whole-number ratios of Pythagorean tradition and the dialectical architecture of Plato's philosophy—to organize inquiry across different domains like music, cosmology, and ethics. The analysis points out the limitations of their approach, including systematic exclusions and a failure to account for phenomena like irrational numbers.

"Sapere Vedere"
An Italian phrase meaning "to know how to see," which Leonardo da Vinci used as a methodological principle. This concept involved a form of trained perception guided by mathematical understanding, anatomical knowledge, and aesthetic sensitivity. The book highlights this as a key to his empirical-aesthetic coordination, which allowed him to bridge the gap between scientific investigation and artistic representation.

Scope Conditions
The specific boundaries and limitations under which a particular coordination method works effectively. The book uses this concept as a critical analytical tool to evaluate the historical figures, arguing that their apparent successes were often contingent on the particular institutional and cultural constraints of their time. The emphasis on scope conditions prevents the report from anachronistically attributing universal principles to historically situated experiments.

Synchronicity
A concept developed by Carl Gustav Jung that refers to "meaningful coincidences"

that lack a causal connection. The book critiques this as Jung's "most problematic theoretical innovation," arguing that the phenomenon can be more parsimoniously explained by cognitive biases, such as confirmation bias and pattern detection bias, rather than by a metaphysical, acausal connecting principle. This analysis provides a clear example of the need for methodological rigor and the importance of separating a compelling subjective experience from a generalized explanatory claim.

Teilhard de Chardin (Pierre)

A Jesuit priest and paleontologist who attempted evolutionary coordination by proposing a Complexity-Consciousness Hypothesis that sought to reconcile his scientific and religious views. The book's critique of his work, alongside that of Iqbal, centers on what it identifies as the "Orthogenesis Problem," which is the assumption that evolution has an inherent, teleological direction.

The de Broglie-Bohm Theory

A mathematically rigorous interpretation of quantum mechanics developed by David Bohm. The book identifies this as his "most lasting contribution to physics" and highlights its significance for demonstrating that deterministic interpretations of quantum phenomena remain mathematically viable, even if they lack empirical advantages over other interpretations. This work is explicitly distinguished from Bohm's later philosophical speculations, serving as a model for separating legitimate scientific contributions from speculative extensions.

Translation as Coordination

A term used to describe the coordination strategy employed by Avicenna and al-Biruni, who built a new philosophical lexicon from existing traditions.1 The book notes that this process was not a literal rendering but a "creative coordination" that transformed both the source and target frameworks, enabling novel theoretical developments.1 The selective appropriation of Greek and Indian texts, and the rejection of others, demonstrates that translation is a strategic, and not neutral, act of intellectual coordination.

.

REFERENCES

Aczel, A. D. (2007). The Jesuit and the skull: Teilhard de Chardin, evolution, and the search for Peking Man. Riverhead Books.

Adami, C. (2016). What is information? Philosophical Transactions of the Royal Society A, 374(2063), 20150230.

Adamson, P. (2016). Philosophy in the Islamic world: A very short introduction. Oxford University Press. Aristotle. (1984). Metaphysics (W. D. Ross, Trans.). In J. Barnes (Ed.), The complete works of Aristotle (Vols. 1–2). Princeton University Press. Barabási, A.-L. (2016). Network science. Cambridge University Press.

Adamson, P. (2016). Philosophy in the Islamic world: A very short introduction. Oxford University Press. Averroes. (1954). The incoherence of the incoherence (S. Van den Bergh, Trans. ; Vols. 1–2). Oxford University Press.

Anderson, P. W. (1972). More is different. Science, 177(4047), 393–396.

Antognazza, M. R. (2009). Leibniz: An intellectual biography. Cambridge University Press.

Aristotle. (1984). *The complete works of Aristotle...* Princeton University Press.

Averroes. (1954). The incoherence of the incoherence. Oxford University Press.

Balaguer, M. (1998). Platonism and anti-Platonism in mathematics. Oxford University Press. Barabási, A.-L. (2016). Network science. Cambridge University Press.

Bargh, J. A., & Chartrand, T. L. (1999). The unbearable automaticity of being. American Psychologist, 54(7), 462-479.

Barrett, J. L. (2004). Why would anyone believe in God? AltaMira Press.

Baxandall, M. (1972). Painting and experience in fifteenth-century Italy. Oxford University Press.

Beck, S., & Mahony, M. (2017). The IPCC and the politics of anticipation. Nature Climate Change, 7(5), 311–313.

Benson, D. A., Cavanaugh, M., Clark, K., Karsch-Mizrachi, I., Lipman, D. J., Ostell, J., & Sayers, E. W. (2018). GenBank. Nucleic Acids Research, 46(D1), D41–D47.

Blei, D. M., Ng, A. Y., & Jordan, M. I. (2003). Latent Dirichlet allocation. Journal of Machine Learning Research, 3, 993–1022.

Bohm, D. (1952a). A suggested interpretation of the quantum theory in terms of "hidden" variables. I. Physical Review, 85(2), 166-179.

Bohm, D. (1952b). A suggested interpretation of the quantum theory in terms of "hidden" variables. II. Physical Review, 85(2), 180-193.

Bohm, D. (1996). On dialogue. Routledge.

Bojinski, S., Verstraete, M., Peterson, T. C., Richter, C., Simmons, A., & Zemp, M. (2014). The concept of Essential Climate Variables in support of climate research, applications, and policy. Bulletin of the American Meteorological Society, 95(9), 1431–1443. (n.d.-a).

Bojinski, S., Verstraete, M., Peterson, T. C., Richter, C., Simmons, A., & Zemp, M. (2014). The concept of essential climate variables in support of climate research, applications, and policy. Bulletin of the American Meteorological Society, 95(9), 1431–1443. (n.d.-b).

Bowker, G. C., & Star, S. L. (1999). Sorting things out: Classification and its consequences. MIT Press.

Boyd, R., & Richerson, P. J. (1985). Culture and the evolutionary process. University of Chicago Press.

Boyer, P. (2001). Religion explained: The evolutionary origins of religious thought. Basic Books. Dürr, D., Goldstein, S., & Zanghì, N. (1992). Quantum equilibrium and the origin of absolute uncertainty. Journal of Statistical Physics, 67(5-6), 843-907.

Bromham, L., Dinnage, R., & Hua, X. (2016). Interdisciplinary research has consistently lower funding success. Nature, 534(7609), 684–687.

Bullard, R. D., & Wright, B. (2009). Race, place, and environmental justice after Hurricane Katrina. Westview Press.

Burkert, W. (2020). Lore and science in ancient Pythagoreanism (Rev. Ed.). Harvard University Press.

Capra, F. (2020). The science of Leonardo: Inside the mind of the great genius of the Renaissance (Rev. Ed.). Anchor Books.

Chakrabarty, D. (2009). The climate of history: Four theses. Critical Inquiry, 35(2), 197–222.

Chalmers, D. J. (1995). Facing up to the problem of consciousness. Journal of Consciousness Studies, 2(3), 200–219.

Cross, N. (2011). Design thinking: Understanding how designers think and work. Bloomsbury Academic.

Davidson, H. A. (1992). Alfarabi, Avicenna, and Averroes on intellect: Their cosmologies, theories of the active intellect, and theories of human intellect. Oxford University Press.

Davies, B. (2021). Thomas Aquinas's Summa Theologiae: A guide and commentary. Oxford University Press.

Deacon, T. W. (2011). Incomplete nature: How mind emerged from matter. W. W. Norton & Company.

Deane-Drummond, C. (2022). Teilhard de Chardin's evolutionary natural theology: A critical appraisal. Zygon, 57(1), 156–178.

Dehaene, S. (2014). Consciousness and the brain: Deciphering how the brain codes our thoughts. Viking.

Dennett, D. C. (1995). Darwin's dangerous idea: Evolution and the meanings of life. Simon & Schuster.

Dorst, K. (2011). The core of 'design thinking' and its application. Design Studies,

32(6), 521–532.

Eccleston-Turner, M., & Upton, H. (2021). International collaboration to ensure equitable access to vaccines for COVID-19: The ACT-Accelerator and COVAX facility. The Milbank Quarterly, 99(2), 426–449.

Edwards, P. N. (2010). A vast machine: Computer models, climate data, and the politics of global warming. MIT Press.

Fine, G. (2019). The possibility of inquiry: Meno's paradox from Socrates to Sextus. Oxford University Press.

Ford, J. D., Pearce, T., Duerden, F., Furgal, C., & Smit, B. (2010). Climate change policy responses for Canada's Inuit population: The importance of and challenges in engaging traditional knowledge. Climatic Change, 102(3–4), 261–282.

Fourcade, M., Ollion, E., & Algan, Y. (2015). The superiority of economists. Journal of Economic Perspectives, 29(1), 89–114.

French, S. (2023). There are no such things as theories: Structural realism and scientific ontology. Oxford University Press.

Fricker, M. (2007). Epistemic injustice: Power and the ethics of knowing. Oxford University Press.

Futuyma, D. J., & Kirkpatrick, M. (2017). Evolution (4th ed.). Sinauer Associates.

Garber, D. (2009). Leibniz: Body, substance, monad. Oxford University Press.

Geertz, C. (1973). The interpretation of cultures. Basic Books.

Giles, J. (2020). The sprint to solve coronavirus protein structures—And disarm them with drugs. Nature, 581(7806), 252–255.

Goldstein, S. (2013). Bohmian mechanics. In E. N. Zalta (Ed.), The Stanford encyclopedia of philosophy (Spring 2013 ed.). Stanford University.

Gould, S. J. (1996). Full house: The spread of excellence from Plato to Darwin. Harmony Books.

Gould, S. J. (1999). Rocks of ages: Science and religion in the fullness of life. Random House.

Grant, E. (1996). The foundations of modern science in the Middle Ages. Cambridge University Press.

Greenwald, A. G., & Banaji, M. R. (1995). Implicit social cognition: Attitudes, self-esteem, and stereotypes. Psychological Review, 102(1), 4-27.

Griffel, F. (2009). Al-Ghazali's philosophical theology. Oxford University Press.

Grosberg, R. K., & Strathmann, R. R. (2007). The evolution of multicellularity: A minor major transition? Annual Review of Ecology, Evolution, and Systematics, 38, 621–654.

Gutas, D. (2022). Greek thought, Arabic culture: The Graeco-Arabic translation movement (2nd ed.). Routledge.

Haidt, J. (2001). The emotional dog and its rational tail: A social intuitionist approach to moral judgment. Psychological Review, 108(4), 814-834.

Hanahan, D., & Weinberg, R. A. (2011). Hallmarks of cancer: The next generation. Cell, 144(5), 646–674.

Harding, S. (2015). Objectivity and diversity: Another logic of scientific research. University of Chicago Press.

Haught, J. F. (2017). The new cosmic story: Inside our awakening universe. Yale University Press.

Haught, J. F. (2021). Teilhard's anticipatory vision: A reading for our times. Teilhard Review, 56(2), 12–29.

Henrich, J. (2016). The secret of our success: How culture is driving human evolution. Princeton University Press.

Holland, P. R. (1993). The quantum theory of motion: An account of the de Broglie-Bohm causal interpretation of quantum mechanics. Cambridge University Press.

Huffman, C. A. (2019). Philolaus of Croton: Pythagorean and Presocratic (2nd ed.). Cambridge University Press.

Hulme, M. (2009). Why we disagree about climate change. Cambridge University Press. Karikó, K., Buckstein, M., Ni, H., & Weissman, D. (2005). Suppression of RNA recognition by Toll-like receptors: The impact of nucleoside modification and the evolutionary origin of RNA. Immunity, 23(2), 165–175.

Ioannidis, J. P. A. (2005). Why most published research findings are false. PLoS Medicine, 2(8), e124.

Iqbal, M. (2013). The reconstruction of religious thought in Islam. Stanford University Press. (Original work published 1930).

Janos, D. (2020). Avicenna on the foundations of metaphysics. Brill.

Jung, C. G. (1921). Psychological types. Princeton University Press.

Jung, C. G. (1952). Synchronicity: An acausal connecting principle. In The collected works of C. G. Jung (Vol. 8). Princeton University Press.

Jung, C. G. (1960). The transcendent function. In The structure and dynamics of the psyche (pp. 67-91). Princeton University Press. (Original work published 1916).

Kellert, S. H., Longino, H. E., & Waters, C. K. (Eds.). (2006). Scientific pluralism. University of Minnesota Press.

Kemp, M. (2019). Leonardo da Vinci: The marvellous works of nature and man (Rev. Ed.). Oxford University Press.

Kitcher, P. (1989). Explanatory unification and the causal structure of the world. In P. Kitcher & W. C. Salmon (Eds.), Scientific explanation (Minnesota Studies in the Philosophy of Science, Vol. 13, pp. 410–505). University of Minnesota Press.

Koshul, B. B. (2014). Muhammad Iqbal's reconstruction of the philosophical argument for the existence of God. Islamic Studies Institute.

Kragh, H. (1999). Quantum generations: A history of physics in the twentieth century. Princeton University Press.

Lamoreaux, G. (2016). Early Eastern Christian contributions to philosophy and science: Hunayn ibn Ishaq and his circle. Intellectual History of the Islamicate

World, 4(1–2), 1–30.

Leahey, E., Beckman, C. M., & Stanko, T. L. (2017). Prominent but less productive? The impact of interdisciplinarity on scientists' research. Administrative Science Quarterly, 62(1), 105–139.

Leibniz, G. W. (1989). Philosophical essays (R. Ariew & D. Garber, Trans.). Hackett Publishing.

Levin, S. A. (1998). Ecosystems and the biosphere as complex adaptive systems. Ecosystems, 1(5), 431–436.

Lindberg, D. C. (2007). The beginnings of Western science (2nd ed.). University of Chicago Press.

Lizzini, O. (2021). Avicenna's metaphysics in context. Edinburgh University Press.

Lloyd, G. E. R. (2021). The ambivalences of rationality: Ancient and modern cross-cultural explorations. Cambridge University Press.

Longino, H. E. (2002). The fate of knowledge. Princeton University Press.

Mach, K. J., Mastrandrea, M. D., Freeman, P. T., & Field, C. B. (2017). Unleashing expert judgment in assessment. Global Environmental Change, 44, 1–14.

Main, R. (2007). Revelations of chance: Synchronicity as spiritual experience. State University of New York Press.

Makdisi, G. (1981). The rise of colleges: Institutions of learning in Islam and the West. Edinburgh University Press.

Malone, T. W. (2018). Superminds: The surprising power of people and computers thinking together. Little, Brown Spark.

McCutcheon, J. P., & Moran, N. A. (2012). Extreme genome reduction in symbiotic bacteria. Nature Reviews Microbiology, 10(1), 13–26.

McGinnis, J. (2020). Avicenna (2nd ed.). Oxford University Press.

McGrath, A. E. (2004). The science of God: An introduction to scientific theology. Eerdmans.

McKirahan, R. D. (2020). Philosophy before Socrates: An introduction with texts and commentary (3rd ed.). Hackett Publishing.

McShea, D. W., & Brandon, R. N. (2010). Biology's first law: The tendency for diversity and complexity to increase in evolutionary systems. University of Chicago Press.

Meier, C. A. (Ed.). (2001). Atom and archetype: The Pauli/Jung letters, 1932-1958. Princeton University Press.

Mesoudi, A. (2011). Cultural evolution: How Darwinian theory can explain human culture and synthesize the social sciences. University of Chicago Press.

Mirowski, P. (2011). Science-mart: Privatizing American science. Harvard University Press.

Mitchell, M. (2009). Complexity: A guided tour. Oxford University Press.

Mitchell, M. (2009). Complexity: A guided tour. Oxford University Press. Plato.

(1997a). Philebus (D. Frede, Trans.). In J. M. Cooper (Ed.), Plato: Complete works (pp. 398–456). Hackett Publishing. Plato. (1997b). Phaedrus (A. Nehamas & P. Woodruff, Trans.). In J. M. Cooper (Ed.), Plato: Complete works (pp. 506–556). Hackett Publishing. Plato. (1997c). Republic (G. M. A. Grube, Trans.; C. D. C. Reeve, Rev.). In J. M. Cooper (Ed.), Plato: Complete works (pp. 971–1223). Hackett Publishing. Plato. (1997d). Sophist (N. White, Trans.). In J. M. Cooper (Ed.), Plato: Complete works (pp. 235–293). Hackett Publishing. Plato. (1997e). Timaeus (D. J. Zeyl, Trans.). In J. M. Cooper (Ed.), Plato: Complete works (pp. 1224–1291). Hackett Publishing.

Mitchell, S. D. (2009). Unsimple truths: Science, complexity, and policy. University of Chicago Press.

Mulchahey, M. M. (2019). "First the bow is bent in study": Dominican education before 1350. Pontifical Institute of Mediaeval Studies. O'Malley, C. D., & Saunders, J. B. de C. M. (1952). Leonardo on the human body: The anatomical, physiological, and embryological drawings of Leonardo da Vinci. Dover Publications.

Nasr, S. H. (1989). Knowledge and the sacred. State University of New York Press.

Nickerson, R. S. (1998). Confirmation bias: A ubiquitous phenomenon in many guises. Review of General Psychology, 2(2), 175-220.

Nordhaus, W. D. (2017). Revisiting the social cost of carbon. Proceedings of the National Academy of Sciences, 114(7), 1518–1523.

Page, S. E. (2011). Diversity and complexity. Princeton University Press.

Pardi, N., Hogan, M. J., Porter, F. W., & Weissman, D. (2018). mRNA vaccines—A new era in vaccinology. Nature Reviews Drug Discovery, 17(4), 261–279.

Park, K. (2006). Secrets of women: Gender, generation, and the origins of human dissection. Zone Books.

Plato. (1997b). *Phaedrus* (A. Nehamas & P. Woodruff, Trans.). Hackett Publishing.

Popper, K. R. (2011). The open society and its enemies (New ed.). Routledge.

Pormann, P. E. , & Savage-Smith, E. (2021). Medieval Islamic medicine (2nd ed.). Edinburgh University Press.

Rahman, F. (1982). Islam and modernity: Transformation of an intellectual tradition. University of Chicago Press.

Richards, R. J. (2002). The romantic conception of life: Science and philosophy in the age of Goethe. University of Chicago Press.

Russell, B. (1937). A critical exposition of the philosophy of Leibniz. Cambridge University Press.

Sachau, E. C. (1910). Alberuni's India: An account of the religion, philosophy, literature, geography, chronology, astronomy, customs, laws and astrology of India (Vols. 1–2). Kegan Paul, Trench, Trübner & Co.

Saliba, G. (2007). Islamic science and the making of the European Renaissance. MIT Press.

Scheid, V. (2002). Chinese medicine in contemporary China: Plurality and synthesis.

Duke University Press.

Seamon, D., & Zajonc, A. (Eds.). (1998). Goethe's way of science: A phenomenology of nature. SUNY Press.

Sepper, D. L. (1988). Goethe contra Newton: Polemics and the project for a new science of color. Cambridge University Press.

Shamdasani, S. (2003). Jung and the making of modern psychology: The dream of a science. Cambridge University Press.

Shapin, S. (1996). The scientific revolution. University of Chicago Press.

Sharp, P. A., & Langer, R. (2011). Promoting convergence in biomedical science. Science, 333(6042), 527.

Simpson, G. G. (1944). Tempo and mode in evolution. Columbia University Press.

Skinner, Q. (1969). Meaning and understanding in the history of ideas. History and Theory, 8(1), 3–53.

Smith, L. T. (2012). Decolonizing methodologies: Research and indigenous peoples (2nd ed.). Zed Books.

Star, S. L., & Griesemer, J. R. (1989). Institutional ecology, "translations" and boundary objects: Amateurs and professionals in Berkeley's Museum of Vertebrate Zoology, 1907–1939. Social Studies of Science, 19(3), 387–420.

Stern, N. (2007). The economics of climate change: The Stern review. Cambridge University Press.

Stump, E. (2003). Aquinas. Routledge.

Tegmark, M. (2000). Importance of quantum decoherence in brain processes. Physical Review E, 61(4), 4194-4206.

Teilhard de Chardin, P. (2008). The phenomenon of man (B. Wall, Trans.). Harper Perennial. (Original work published 1959).

Thompson, E. (2007). Mind in life: Biology, phenomenology, and the sciences of mind. Harvard University Press.

Thurner, S., Hanel, R., & Klimek, P. (2018). Introduction to the theory of complex systems. Oxford University Press.

Tononi, G., & Koch, C. (2015). Consciousness: Here, there and everywhere? Philosophical Transactions of the Royal Society B, 370(1668), 20140167.

Tononi, G. (2008). Consciousness as integrated information. Biological Bulletin, 215(3), 216-242.

Tversky, A., & Kahneman, D. (1973). Availability: A heuristic for judging frequency and probability. Cognitive Psychology, 5(2), 207-232.

Varela, F. J., Thompson, E., & Rosch, E. (1991). The embodied mind: Cognitive science and human experience. MIT Press.

Vincent, J. F. V., Bogatyreva, O. A., Bogatyrev, N. R., Bowyer, A., & Pahl, A.-K. (2006). Biomimetics: Its practice and theory. Journal of the Royal Society Interface, 3(9), 471–482.

Vogelstein, B., Papadopoulos, N., Velculescu, V. E., Zhou, S., Diaz, L. A., & Kinzler, K. W. (2013). Cancer genome landscapes. Science, 339(6127), 1546–1558.

Wasserstein, R. L., & Lazar, N. A. (2016). The ASA statement on p-values: Context, process, and purpose. The American Statistician, 70(2), 129–133.

Watson, J. D., & Crick, F. H. C. (1953). Molecular structure of nucleic acids: A structure for deoxyribose nucleic acid. Nature, 171(4356), 737–738.

West, G. (2017). Scale: The universal laws of growth, innovation, sustainability, and the pace of life. Penguin Press.

Westfall, R. S. (1980). Never at rest: A biography of Isaac Newton. Cambridge University Press.

Whitehead, A. N. (1978). Process and reality (Corrected ed.). Free Press.

Whitson, J. A., & Galinsky, A. D. (2008). Lacking control increases illusory pattern perception. Science, 322(5898), 115-117.

Whyte, K. P. (2017). Indigenous climate change studies: Indigenizing futures, decolonizing the Anthropocene. English Language Notes, 55(1), 153–162.

Wiesner-Hanks, M. E. (2019). Women and gender in early modern Europe (4th ed.). Cambridge University Press.

Wilson, E. O. (1998). Consilience: The unity of knowledge. Knopf.

Wuchty, S., Jones, B. F., & Uzzi, B. (2007). The increasing dominance of teams in production of knowledge. Science, 316(5827), 1036–1039.

Wulf, A. (2015). The invention of nature: Alexander von Humboldt's new world. Knopf.

Zhmud, L. (2022). Pythagoras and the early Pythagoreans (Rev. Ed.). Oxford University Press.

www.ingramcontent.com/pod-product-compliance
Lightning Source LLC
Chambersburg PA
CBHW071434210326

41597CB00020B/3781